Internet of Things in Modern Computing

The text focuses on the theory, design, and implementation of the Internet of Things (IoT), in a modern communication system. It will be useful to senior undergraduate, graduate students, and researchers in diverse fields domains, including electrical engineering, electronics and communications engineering, computer engineering, and information technology.

Features:

- Presents all the necessary information on the Internet of Things in modern computing
- Examines antenna integration challenges and constraints in the Internet of Things devices
- Discusses advanced Internet of Things networks and advanced controllers required for modern architecture
- Explores security and privacy challenges for the Internet of Things-based healthcare system
- Covers implementation of Internet of Things security protocols such as MQTT, Advanced Message Queuing Protocol, XMPP, and DSS

The text addresses the issues and challenges in implementing communication and security protocols for the IoT in modern computing. It further highlights the applications of IoT in diverse areas, including remote health monitoring, remote monitoring of vehicle data and environmental characteristics, Industry 4.0, 5G communications, and Next-gen IoT networks. The text presents case studies on IoT in modern digital computing. It will serve as an ideal reference text for senior undergraduate, graduate students, and academic researchers in diverse fields domains, including electrical engineering, electronics and communications engineering, computer engineering, and information technology.

Smart Technologies for Engineers and Scientists
Series Editor: Mangey Ram

Applications of Mathematical Modeling, Machine Learning, and Intelligent Computing for Industrial Development
Madhu Jain, Dinesh K. Sharma, Rakhee Kulshrestha, and H.S. Hota

Internet of Things in Modern Computing
Theory and Applications

Edited by
Vinay Chowdary
Abhinav Sharma
Naveen Kumar
Vivek Kaundal

CRC Press
Taylor & Francis Group
Boca Raton London New York

CRC Press is an imprint of the
Taylor & Francis Group, an **informa** business

First edition published 2023
by CRC Press
6000 Broken Sound Parkway NW, Suite 300, Boca Raton, FL 33487-2742

and by CRC Press
4 Park Square, Milton Park, Abingdon, Oxon, OX14 4RN

CRC Press is an imprint of Taylor & Francis Group, LLC

© 2024 selection and editorial matter, Vinay Chowdary, Abhinav Sharma, Naveen Kumar and Vivek Kaundal; individual chapters, the contributors

ISBN: 978-1-032-39272-1 (hbk)
ISBN: 978-1-032-52582-2 (pbk)
ISBN: 978-1-003-40730-0 (ebk)

DOI: 10.1201/9781003407300

Typeset in Sabon
by SPi Technologies India Pvt Ltd (Straive)

Contents

Preface

The Internet of Things (IoT) has emerged as the major contributor behind the development of the technological infrastructure which connects physical objects to the internet. It is evident that an examination of the role of the IoT in the modern era is urgently needed. Therefore, the intention of this publication is to have all the necessary information regarding the IoT in modern computing contained within a single cover.

The organization of the book is as follows:

Chapter 1 lays the foundation stone in terms of the progress made in the Internet of Things (IoT) field and advance the challenges in the architecture design of IoT. The chapter also discusses both classical and emerging IoT models.

Chapter 2 provides a bird's-eye view of Industry 4.0 and the role of Artificial Intelligence (AI) and the Internet of Things (IoT) in achieving this new framework. It provides a quick overview of the history of the industrial revolution by highlighting some significant historical events. Industry 4.0 is a new model that strives to build an open and smart processing platform.

Chapter 3 discusses the role of edge computing and its ability to do sophisticated calculations at the edge location rather than doing the calculations at the cloud location. In this chapter recent study, developments and benefits in edge computing are also highlighted. There are a few illustrations and photographs in this chapter.

Chapter 4 discusses the role of Industrial IoT (IIoT) within Industry 4.0. IIoT, which is an integral part of IoT, is one of the most widely used terms in the development of modern computing. This chapter offers some detailed insights with regard to IIoT, along with an outline of its key technologies. The chapter also focuses on the open challenges and application domain of IoT in Industry 4.0.

Chapter 5 provides a general framework for the detection of Denial of Service (DoS) attacks in the field of IoT. It analyses the research gaps in DoS detection techniques and then offers a two-stage framework for DoS detection technique based on the Fuzzy Rule Manager and Neural Networks.

Chapter 6 examines the parameters that may either accelerate or hinder the adaption of IoT by means of a skill development course. The indicated study extended the Unified Theory of Acceptance and Use of Technology model through an integration of added factors, i.e., computer self-efficacy along with relative advantages.

Chapter 7 highlights the key insights of different methods of implementing IoT, AI and blockchain in the healthcare sector and also their medical applications in disease diagnosis and treatment, sensor networks, and some future aspects regarding data privacy, security, and safety.

Chapter 8 presents an overview of the coordination issues faced in IoT-driven Building Energy Management Systems (BEMS) subsystems owing to their heterogeneous nature and how interoperability tries to address the same. It also emphasizes the need for a decision support model in the measurement and management of energy usage in buildings.

Chapter 9 focuses on the role of the smart city of the future in pandemic control. Wireless sensor networks and the Internet of Things are the backbone of a smart city. The chapter authors have simulated the case of a smart city using the CupCarbon™ tool. The authors traced the movement of infectious nodes and the spread has been analysed.

Chapter 10 highlights the applications of the IoT in the field of robotics. The authors propose a remotely operated Mobile Manipulator that can detect suspicious or explosive devices both during terrorist attacks and on active bomb sites. The robot designed is operated wirelessly from a remote location in order to make the bomb disposal safer.

Chapter 11 addresses a need for the IoT in agricultural areas in order to reduce the use of chemical fertilizers and crop protection agents. In this study the IoT devices are designated to be used for data collection and to identify the pests in the photos. The chapter intends to create a model for pest identification and classification.

Chapter 12 proposes a framework to utilize the On-Board Diagnostic-II (OBD II) module to collect vehicle data and a smartphone for multiple inbuilt sensors and connectivity employing the IoT cloud. Different data cleaning, characterization, machine learning and analytics models are used to present and framework and the results.

Chapter 13 proposes a printed monopole radiator-based two-element array for the Internet of Things (IoT)-based applications and commercial 5G (3.3–3.8 GHz) connectivity. The design proposed in this chapter is suitable commercial 5G communications and IoT-based applications.

Chapter 14 proposes an automated system for the identification of dental diseases using deep learning and the IoT. The technique proposed in this study is extremely efficient and has a very low probability of error. The experimental results of the proposed technique are also provided.

This research book is intended to cater to the requirements of a wide domain of researchers working in the domain of AI, robotics and IoT/WSN. This book covers a basic understanding of IoT and its architecture design along with the application of advanced technologies such as AI, Industrial

IoT, deep learning and 5G. This book serves not only academic and technical readers but also those based in industry, allowing them to gain some insights into the latest developments in the technological framework and its impact. Practical case studies from leading industries such as Bosch, Larsen & Toubro, etc are also a part of this book. This book can also serve as a valuable reference for academics, mechanical, mechatronics, computer science, information technology and industrial engineers, and environmental scientists as well as researchers in related subjects.

Dehradun, India

Editors:
Vinay Chowdary
Abhinav Sharma
Naveen Kumar
Vivek Kaundal

Acknowledgments

The editor acknowledges CRC Press - Taylor & Francis Group for this opportunity and professional support. My special thanks to Gagandeep Singh, Publisher Engineering, CRC Press, Gauravjeet Singh Reen, Senior Editor - Engineering and Isha Ahuja, Editorial Assistant - Engineering for the excellent support provided to us during the completion of this book.

Thanks to the chapter authors and reviewers for their availability for this work.

Vinay Chowdary, International School of Engineering, India
Abhinav Sharma, University of Petroleum and Energy Studies, India
Naveen Kumar, Christ University, India
Vivek Kaundal, Larsen & Toubro Technology Services, India

Editors

Prof. Dr. Vinay Chowdary received his B.Tech. M.Tech. degrees from MIST, JNTUH, Hyderabad, India, in 2005 and 2012, respectively, and a Ph.D. degree from UPES, Dehradun, India, in 2020. He is currently working as an Associate Professor at the uGDX Institute of Technology, India. He has more than 14 years of teaching and research experience and extensive experience in curriculum design. His research interests include embedded systems, the Internet of things, wireless sensor networks, and advanced microcontrollers.

Prof. Dr. Abhinav Sharma received his B.Tech. from H. N. B. Garhwal University, Srinagar, India, in 2009 and his M.Tech. from Govind Ballabh Pant University of Agriculture and Technology, Pantnagar, India, in 2011. He completed his Ph.D. from Govind Ballabh Pant University of Agriculture and Technology, Pantnagar, India, in 2016. He is currently working as an Assistant Professor (Senior Scale) in the Department of Electrical and Electronics Engineering in University of Petroleum and Energy Studies, Dehradun. His fields of interest includes adaptive array signal processing, artificial intelligence, machine learning, optimization techniques and smart antennas.

Dr. Naveen Kumar received a B.Tech. from SVIET, Punjab, India in 2009 and an M.E. from NITTTR, Chandigarh, India in 2013. He is currently working as a Research Engineer, FCS Railenium, France where the design and development of antennas for railways is being carried out under the European project 'Shift2Rail'. He completed his Ph.D. in July 2021 from Thapar Institute of Engineering and Technology (Deemed to be University), Punjab, India. He has more than six years of experience in research, academics, and corporate domains. He is the inventor of a novel antenna for which a French patent was filed in June 2020. He has published almost 50 papers in various reputable national/international journals and conferences. He is an active member of IEEE, EuMA, and IAENG. His research interests are antenna design, antennas for IoT, MIMO antennas, metamaterials, and characteristic mode analysis.

Dr. Vivek Kaundal received his B.Tech in Electronics & Communication Engineering, M.Tech. in Digital communication, and a Ph.D. in Wireless Sensor Networks (Electronics Engineering). He has more than 13 years of wide-ranging experience in the corporate sector and academia. He is currently working for L&T Technology Services as a Specialist in LTTS, Mysore. In this position he is responsible for designing and developing a competency framework in alignment with business needs. His field of interests are automotive and embedded systems. He has a rich experience in AUTOSAR, C4K tool, and Mentor Graphics (CVI) tool for the configuration of AUTOSAR modules like CAN, Diagnostics, OS etc. and in integration. He has also worked in the area of application software components, root software composition file etc.

Contributors

KNS Acharya
L&T Technology Services
Limited
Bangalore, India

K. S. Ashwini
Coventry University
United Kingdom

Chandra Shekar Besta
National Institute of Technology
Calicut, Kerala, India

Manoj Bhatt
College of Technology,
   G.B.P.U.A&T.
Pantnagar, Uttarakhand, India

P. S. Birla
FOE, IGNTU
Amarkantak, Madhya Pradesh,
   India

Asmita Singh Bisen
Sharda University
Greater Noida, Uttar Pradesh, India

Chandrashekar
L & T Technological Services
Bangalore, India

Anirban Chatterjee
National Institute of Technology
Farmagudi, Ponda, Goa

Dinesh Chaudhari
JDIET
Yavatmal, Maharashtra, India

Vinay Chowdary
uGDX Institute of Technology
Hyderabad, India

Parul Dubey
G H Raisoni College of
   Engineering
Nagpur, India

Pushkar Dubey
Pandit Sundar Lal Sharma (Open)
   University
Chhattisgarh, India

Shival Dubey
Institute of Design, Robotics &
   Optimization, University of
   Leeds
United Kingdom

Ankit Gupta
University of Petroleum & Energy
   Studies
Dehradun, Uttarakhand, India

Mukul Kumar Gupta
University of Petroleum and Energy
   Studies
Dehradun, Uttarakhand, India

Neeraj Gupta
K.R. Mangalam University Panipat
   Institute of Engineering and
   Technology
Panipat, Haryana, India

Avinash P. Jadhav
DRGIT&R Amravati
Maharashtra, India

Samyak Jain
University of Petroleum & Energy
   Studies
Dehradun, Uttarakhand, India

Deepak Chandra Joshi
Shiv Nadar University
Dadri, Ghaziabad, India

Abhishek Joshi
Sungkyunkwan University
Seoul, South Korea

Atul B. Kathole
D. Y. Patil Institute of Technology
Pune, Maharashtra, India

Kalyan Sundar Kola
National Institute of Technology
Farmagudi, Ponda, Goa

Sanjay Mathur
College of Technology,
   G.B.P.U.A&T
Pantnagar, Uttarakhand, India

Sindhu P. Menon
Professor, School of Computing and
   Information Technology, REVA
   University
Bangalore, Karnataka, India

Pramod Kumar Naik
Associate Professor, Department of
   Computer Science and Engineering,
   Dayananda Sagar University
Bangalore, Karnataka, India

Ananta Narayana
DHSS, MNNIT
Allahabad, Uttar Pradesh, India

Prithvi Sekhar Pagala
L&T Technology Services Limited
Bangalore, India

Sonali D. Patil
Pimpri Chinchwad College of
   Engineering
Pune, Maharashtra, India

Aditi Paul
Banasthali Vidyapith
Rajasthan, India

Himanshu Payal
Sharda University
Greater Noida, Uttar Pradesh, India

Thinagaran Perumal
University of Putra
Selangor, Malaysia

Suhailam P
National Institute of Technology
Calicut, Kerala, India

Kailash Kumar Sahu
Pandit Sundar Lal Sharma (Open)
   University
Chhattisgarh, India

Abhinav Sharma
University of Petroleum & Energy
   Studies
Dehradun, Uttarakhand, India

Thrilochan Sharma
L. & T. Technology Services
Limited
Bangalore, India

S Sinha
CHRIST (Deemed to be University)
Bangalore, Karnataka, India

R. K. Shastri
DHSS, MNNIT
Allahabad, Uttar Pradesh, India

Sumukh Surya
Bosch Global Software
Technologies
Bangalore, India

Mohan Krishna S.
Alliance University
Bangalore, India

Baskar Venugopalan
Professor, Department of Computer
  Science and Technology,
  Dayananda Sagar University
Bangalore, Karnataka, India

Rashi Verma
National Institute of Technology
Calicut, Kerala, India

Kapil N. Vhatkar
Pimpri Chinchwad College of
  Engineering
Pune, Maharashtra, India

Saumya Yadav
Indraprastha Institute of
  Information Technology (IIIT)
Delhi, India

Raju Yerolla
National Institute of Technology
Calicut, Kerala, India

Chapter 1

IoT architecture and design

Neeraj Gupta

Panipat Institute of Engineering and Technology, Panipat, India

CONTENTS

1.1 INTRODUCTION

Industry 4.0 refers to the automation of the decision-making process to optimize various usages of resources by enabling interactions between the physical and the digital world. The Internet of Things (IoT) is one of the critical technologies used to facilitate the above vision and the purpose of Industry 4.0. IoT involves the participation of many technologies, including sensors, communication protocols, edge computing, fog computing, data analytics, and cloud computing. The design of the architecture involves specifications of various functional modules and the interaction between these modules to create a system. According to the Gartner glossary, "The Internet of Things (IoT) is the network of physical objects that contain embedded technology to communicate and sense or interact with their internal states or the external environment" [1]. The physical objects are the sensors employed to sense and detect the changes in the physical environment for specific applications. The recorded parameters are then communicated and processed locally or on cloud-based systems in order to make well-informed decisions. IoT is an amalgamation of various technologies, including sensor networks, computer networks, cloud computing, and data analytics. These technologies form the driving force that contributes to the functionality of different IoT layers. Regardless of disruption as a result of chip shortages, and global events, it is estimated that there will be 27 billion IoT devices by the end of the year 2025 [2].

The current chapter is divided into three main sections. The first section discusses the challenges involved in IoT architecture. The second section gives insight into two prevalent classical architecture models of IoT. The third section outlines the current state of the art in order to enhance the capabilities of the system.

1.1.1 Challenges: IoT architecture drivers

The deployment of trillions of sensors to capture data from billions of systems to empower millions of applications have enabled business sectors to make informed and intelligent decisions. Global vendors are now offering customized solutions to facilitate and automize processes across various public and private sectors. However, multiple challenges need to be addressed to ensure that the solutions offered remain meaningful:

1. Scale: The explosive growth of smart things has burdened the network with enormous amounts of data. It is estimated that by 2025 total data volume generated by IoT devices will reach 79.4 zettabytes. The rollout of 5G is expected to further accelerate the growth of connected cars, smart home devices, and industrial equipment. According to forecasts, there will be 30.5 billion IoT devices across the world by 2025. Each of these devices will require the allocation of a unique address in order to communicate. It is essential to strategies IPv6 addresses allotment to meet the growing demand for connected devices.
2. Security: As per the report by [3], the following statistics underline the importance of cybersecurity in the Internet of Things.
   a. Cybercrime is estimated to generate about $10.5 trillion 2025.
   b. 98% of IoT devices are ill-equipped to deal with cyber-attacks.
   c. On average, 97% of encrypted files affected by so-called ransomware attacks are recovered after companies pay the criminals.
      The components in IoT need to be connected in order to share the information. In this process the role of data security is of paramount importance. The basic building blocks of confidentiality, integrity, and availability need to be ensured at various stages, including nodes, gateways, fog, and cloud computing.
3. Privacy: With users' increasing use of smart devices, the generated data is more individual-specific. This may be wide-ranging in nature: financial, health-related, location-based, and/or shopping patterns, to list just a few. The service provider stores and processes this information as one aspect of their offered services. The service provider must clearly state their policies, stating the status of ownership and accessibility of the information with other parties.

4. Interoperability: The solution offered by the various organizations in the IoT domain is based on protocols that are either proprietary solutions or belong to open standards. There are many protocols and architectures, each with their own sets of advantages and disadvantages. The need to support the legacy devices, along with new devices with more outstanding capabilities, calls for the provision of interoperability solutions. It is imperative to standardize architecture at the global level in order to ensure interoperability among solution providers so that the maximum benefits of technology can be reaped.

5. Constrained Nodes and Networks: Wireless Sensor Networks (WSN) are an essential component of IoT. The sensor and sensor networks are usually application-specific. The nodes or sensors are resource-constrained devices with both limited processing capabilities and limited power. At a different layer of IoT architecture, many efficient protocols have been proposed and deployed successfully across the field. Among the most prominent architectures advanced at the MAC layer at present are IEEE 802.15.4 [4], 6LoWPAN [5], HART, and Zigbee [6]. In addition, Message Queue Telemetry Transport Protocol, Constrained Application Protocol, Advanced Message Queueing Protocol, and Extensible Messaging and Presence Protocol are among the application-based protocols.

6. Data Processing and Analysis: The massive amount of sensed and communicated data needs to be processed and analyzed in order to produce meaningful information. Information processing can occur at various points of the process, including sensor nodes, edge/fog elements, and cloud servers. The criticality of the application and volume of data define the stage of the networking element where processing needs to be executed. Data parameters, including velocity, variety, variability, and veracity, need to be considered when processing the data.

7. Heterogeneity: The concept of the IoT promotes the idea of computing anywhere anytime such that interconnected systems support device independence, location independence, and data format independence. To realize the principle of autonomic computing, it is essential to keep both IP-based systems and non-IP-based systems. For example, various devices support IPv4 instead of IPv6. There is a need to develop and design middleware solutions that can overcome multiple heterogeneous issues at multiple levels of architecture design.

1.1.2 Classical IoT architecture models

There are various elements and technologies involved in realizing the essence of IoT. Data collection, data transportation, data analysis, and actuation based on analysis form the basic building blocks for IoT architecture. There are primarily two classical models that are mentioned in the

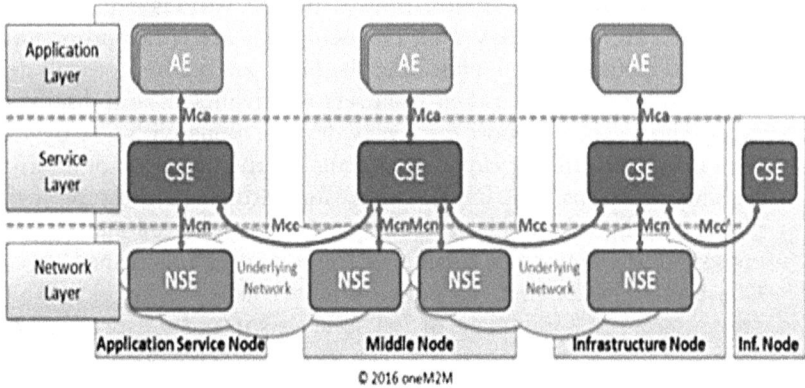

© 2016 oneM2M

Figure 1.1 oneM2M architecture stack [8].

literature: oneM2M architecture (as shown in Figure 1.1) and IoT World Forum Reference Model.

1.1.2.1 oneM2M architecture

This has developed from a project [7] founded by eight prominent ICT standard development organizations in 2012 to address an issue related to the interoperability and scalability issues faced by IoT technologies. The oneM2M standard is open access in nature. The architecture is a middleware technology connecting applications and devices via RESTful APIs. In this system the architecture stack consists primarily of three layers: the Application Layer, the Service Layer, and the Network Layer:

A. Application Layer: The Application Entity (AE) implements the application service logic in the topmost layer. The running instances of the application(s) should have a unique id (AE-ID). These instances can be either virtualized or physical. The communication between the instances is carried out by sending the request to the service layer.

B. Service Layer: The middleware layer contains Common Service Functions (CSF) residing in the service layer. CSF can be present in sensors, gateways, edge devices, and back-end cloud servers. There are 14 CSFs that handle device management, registration, security, semantic interoperability, and location services. Common Service Entities (CSE) represent an instance of a set of CSF. Each CSE is identified by unique CSE-id. The AE invokes the CSE to utilize a particular service.

C. Network Layer: This layer includes the sensors and communication protocols that are used to transfer the data. Network Service Entities provide their service to the CSE. The services being offered by this later include device location, device triggering, and various sleep modes.

Figure 1.2 Horizontal architecture [8].

The term *node* has a particular meaning in oneM2M standards. It is defined in the literature as

> Nodes are logical entities identifiable in the oneM2M System. oneM2M Nodes typically contain CSEs and/or AEs. For the definition of Node types, oneM2M distinguishes between Nodes in the 'Field Domain' – i.e., the domain in which sensors/actuators/aggregators/gateways are deployed – and the 'Infrastructure Domain' – i.e., the domain in which servers and applications on larger computers reside [8].

It is important to mention that the oneM2M standard supports both IP-based and non-IP-based systems through Interworking Proxy Entities (IPE). There are unique AEs that facilitate communication between two incompatible systems to communicate with each other. The horizontal architecture in Figure 1.2 offers various advantages:

1. Reusability by offering common service layer functionality.
2. Interoperability is achieved at various layers using API.
3. The issue of heterogeneity can be addressed seamlessly.

1.1.2.2 IoT world forum reference model

IoT World Forum (IoTWF) is an industry event organized annually by CISCO, as shown in Figure 1.3. In 2014, the IoTWF Consortium unveiled a seven-layer architecture to simplify, clarify, identify, standardize, and organize various elements and functions of IoT systems and applications. The terminologies defined in each layer of the reference model help standardize the component's scope and functionality. The most important feature of the model is the bidirectional flow of data across various layers. The information

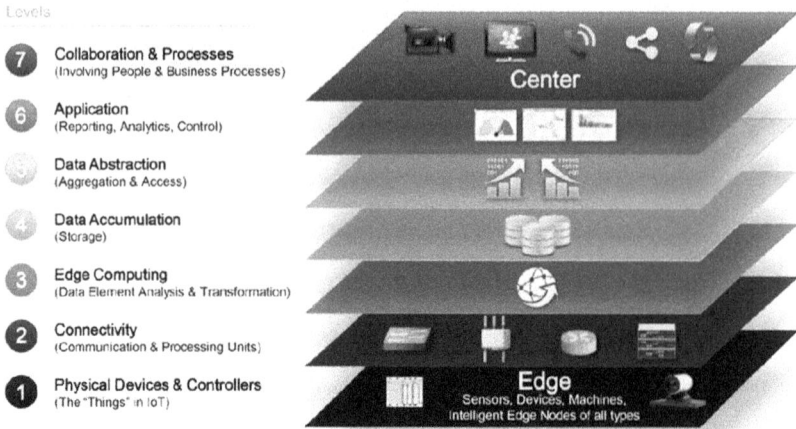

Figure 1.3 IOTWF reference model.

flow can be categorized as either control flow or monitoring flow. In control flow, the information flows from the highest layer to the bottom layer; the process is reversed in monitoring flow:

A. First Layer: This layer describes 'things' in IoT, which can generate data, perform conversion from analog to digital and vice versa, and can be queried and controlled from a remote location. A low level of processing can be quantified in the devices.

B. Second Layer: This layer is concerned with communication and connectivity protocols among the devices. The existing network protocols can be used to transfer data
1. Between devices on the same network.
2. Between devices across different networks
3. To the device/element present in the upper layer.
To enable communication through devices that are not IP-compliant, the model suggests using gateways. The network security and network analytics could also be implemented at this layer.

C. Third Layer: This layer focuses on data analysis and transformation by processing the data. The process can either be used to control the devices based on the type of event generation or be stored at a higher layer for storage and further processing. Edge computing and fog computing are the related terminologies that can be referred to for this layer.

D. Fourth Layer: The fourth layer is concerned with data accumulation and is accessible to applications on a non-real-time basis. The accessibility is converted from event-based need to query-based processing.

E. The accumulated data are further filtered and stored in database tables.

F. Fifth Layer: The data abstraction layer handles the variety, variability, and veracity of data. This layer's responsibility is to ensure that consolidated data stored in the cloud is both reconciled and consistent. The data generated from various sources can have different formats and semantics.

G. Sixth Layer: Various software and algorithms are deployed to interpret the consolidated data based on the need of the business application.

H. Seventh Layer: The IoT involves people and processes. The aim is to collaborate with different processes and people to ease the decision-making. Collaboration and processes are the right terms coined for this layer.

The model recommends that security measures must

1. Secure each device
2. Provide security to process and at each level.
3. Secure communication of data across various layers.

1.1.3 Emerging IoT architecture models

Different applications have their own set of challenges that need customized solutions. Heterogeneous and voluminous data should be appropriately processed to ensure that business decisions can be taken responsibly. There may be a requirement to make fast decisions to drive real-time applications. Customer-oriented applications require transparency and security while transactions are carried out. Different technologies cater to different roles and responsibilities of 'things'. This section categorizes IoT based on a different level of reference model, providing appropriate solutions. The classification is based on key technologies that paved the way for modern computing and numerous applications.

A. Cloud-based IoT
   Cloud computing has revolutionized how computing services, both hardware and software, are being offered. Virtualization has played a vital role in facilitating multi-tenant architecture. The storage facility of cloud service providers accumulates the massive volume of data generated from multiple sources. The cloud can act as a platform for sharing typical applications among numerous 'things' located at various locations. The enormous amount of processing power can be utilized to process for analysis. The disadvantages of the clouds include high latency, security, energy consumption, and billing.

B. Fog Computing-based IoT
   The term fog computing was coined by Cisco Systems to reflect the extension of the services provided by cloud computing to the edge of the network. Fog computing provides immediate processing, storage,

and application services to the physical IoT devices. The aim of introducing the fog layer is to reduce latency and increase the quality of service for the real-time applications to lower reaction time, for example, in industrial applications, healthcare, and transportation. The significant advantages offered are:

1. That it leads to reduced network congestion, thus improving the latency period and lower bandwidth consumption.
2. That it addresses the security and privacy issues on a local level rather than on a global level.
3. That it exhibits a high level of flexibility, scalability, and fault tolerance can be attained.

To supplement the requirements of IoT, fog computing environments are required to address the challenges posed with regard to scalability, interoperability, security, data quality, security, location-awareness, mobility, and reliability. In order to do so, efforts have been made to standardize the fog architecture. Open Edge Computing, Open Fog Computing, and Mobile Edge Computing are among of the major architectural models proposed by academia and industry.

C. SDN-based IoT

The network elements are responsible for forwarding and managing data packets in the networks. In traditional networking, each networking component needs to be handled separately. This separation leads to the cumbersome of managing an extensive network. Software-Defined Networks (SDN) is an emerging architecture that enables the separation of control logic from the networking elements. The controller defines the rules and policies required to manage various networking elements. The logically centralized control provides a global view of the network and aids in framing rules to ensure various complexities are managed well. SDN architecture has three main components: Data Plane/Infrastructure Layer; Control Layer; and Application Layer [9]. The layers interact with each other through an Application Programming Interface (API) based on Representational State Transfer (REST) architecture. In the context of IoT, SDN provides the capability to handle dynamically and control management issues. The primary advantages of this approach are effective management, the seamless handling of mobile devices, the efficient distribution of resources, and optimized energy management [10]. Some challenges require attention from both academia and private industry to fully realize the power of SDN [11].

1. Fault Tolerance: The resiliency of SDN networks is a major issue. There is a need to design and implement the high fault-tolerant network operating systems for controllers.
2. Scalability: The movement of voluminous data consumes resources and constrains the network capabilities. Network Function Virtualization can provide alternate solutions, but the issue of performance analysis based on the size of networks needs to be addressed.

3. Controller Placement: Controllers define the working of SDN elements and control the QoS parameters of the network. Depending on the size of networks and their requirement, it is important to decide the number of controllers, controller location, flow rules, and assignment of devices to the controller.
4. Security: The controller's security from both external and internal networks is of paramount importance. Northbound flow and Southbound flow should undergo a rigorous authentication process to maintain digital security.

D. Blockchain-based IoT

Blockchains enhance data security whenever multiple users want to share the data without losing the control and ownership of it. It is a kind of distributed database containing immutable data stored in blocks. It is important to mention the data is organized using linear data structures like a linked list. In the literature, blockchains are also referred to as Distributed Ledger Technology (DLT). The consensus algorithms, such as Proof of Work, Proof of Stake, and Practical Byzantine Fault Tolerance, ensure the reliability and validity of the transactions done by the users. Industrial IoT, healthcare, and the Internet of Vehicles are among the key applications where blockchains have been applied. The key issues that need attention are [12]:

1. Data Insertion: The data generated by the nodes need to be inserted as a block and verified. There are primarily three ways to operate: client approach, manual approach, and via a connector. The constrained nodes in the client mode insert the partial data. In the manual system, the owner of the 'things' enters the data manually using its blockchain address. The nodes in the fog/edge layer can then take responsibility for adding the data to the chain, which can be another viable option.
2. Identity Model: Most of the literature uses Public Key Infrastructure (PKI) as the default way to uniquely identify the node in the IoT network. Few solutions based on ontology and combinations of hierarchical things and PKI have been proposed, however. There is a need to develop a more scalable solution for identity management.
3. Maintenance: The aspect reflects the security and privacy issues. Forking causes a split in the chain whenever the rules governing the blockchain protocol are changed. This situation can jeopardize the overall security of the system. Similarly, a malicious node's leak of hash function can be a difficult situation to deal with.

Other issues, such as latency, energy, scalability, and flexibility, need to be addressed [13].

E. Analytics-based IoT: Data analytics is essential to summarize the information and knowledge from the data 'things' in the field. This data is voluminous, varied, and sometimes requires fast processing in real time. Most of the received data is unstructured. It is important to undertake pre-processing activity to make it usable for analysis

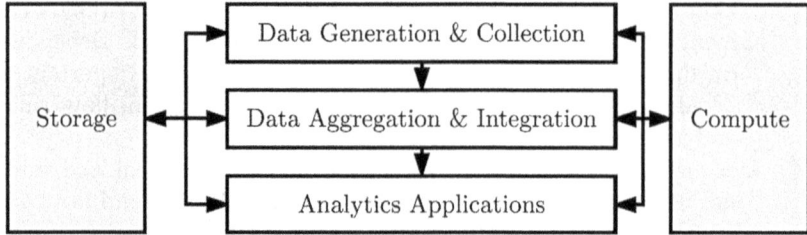

Figure 1.4 Process flow for data analytics [14].

purposes. The analytics can be classified into five categories: descriptive, diagnostic, discovery, predictive, and prescriptive. Figure 1.4 illustrates the process for analytics application

Dealing with a variety of data needs more focused attention. Predictive and prescriptive analytics can be made more accurate by improving the quality of input data.

Architectural models based on mobility [15], Artificial Intelligence [16], information-centric networking [17], and 5G-IoT [18] are all emerging topics that require active research participation from academia and industry.

1.1.4 Conclusion

The Internet of Things has vast potential to address the issues concerning day-to-day human activities. Wireless sensor networks, fog computing, and data analytics form the basic building blocks for IoT. This chapter discussed the basic terminologies, challenges, and key IoT architectural models. Apart from the two classical models that aim to standardize the IoT architecture, emerging paradigms, such as blockchains, VANETS, and Artificial Intelligence, are being incorporated to enhance the capabilities of IoT. The applications based on the IoT facilitate real-time monitoring, increase productivity, and reduce the need for human effort. The COVID era underlines the necessity and usefulness of IoT applications. Future research activities need to focus on the heterogeneity of devices, scalability, security, and energy-efficient systems.

REFERENCES

1. "Definition of Internet Of Things (iot) - IT Glossary | Gartner." https://www.gartner.com/en/information-technology/glossary/internet-of-things (accessed May 01, 2022).
2. "State of IoT 2021: Number of connected IoT devices growing 9% to 12.3 B." https://iot-analytics.com/number-connected-iot-devices/ (accessed May 01, 2022).
3. "73 Important Cybercrime Statistics: 2021/2022 Data Analysis & Projections - Financesonline.com." https://financesonline.com/cybercrime-statistics/ (accessed May 03, 2022).

4. "IEEE SA - IEEE 802.15.4-2020." https://standards.ieee.org/ieee/802.15.4/7029/ (accessed May 04, 2022).

5. "RFC 6282 - Compression Format for IPv6 Datagrams over IEEE 802.15.4-Based Networks." https://datatracker.ietf.org/doc/html/rfc6282 (accessed May 04, 2022).

6. "CSA-IOT - Connectivity Standards Alliance." https://csa-iot.org/ (accessed May 04, 2022).

7. "OneM2M overview - OneM2M." https://wiki.onem2m.org/index.php?title= OneM2M_overview (accessed May 05, 2022).

8. "Using oneM2M." https://onem2m.org/using-onem2m/developers/basics?jjj= 1647576221153 (accessed May 05, 2022).

9. E. Haleplidis, S. Denazis, J. H. Salim, O. Koufopavlou, D. Meyer, and K. Pentikousis, "SDN Layers and Architecture Terminology," 2015. [Online]. Available: http://tools.ietf.org/html/draft-haleplidis-sdnrg-layer-terminology-04.

10. H. Zemrane, Y. Baddi, and A. Hasbi, "SDN-Based Solutions to Improve IOT: Survey," *Proc. – 2018 IEEE 5th International Congress on Information Science and Technology (CiST), Morocco*, pp. 588–593, 2018, doi: 10.1109/CIST.2018. 8596577.

11. I. Deva Priya and S. Silas, *A Survey on Research Challenges and Applications in Empowering the SDN-Based Internet of Things*, vol. 750. Springer Singapore, 2019.

12. S. K. Lo et al., "Analysis of Blockchain Solutions for IoT: A Systematic Literature Review," *IEEE Access*, vol. 7, pp. 58822–58835, 2019, doi: 10.1109/ ACCESS.2019.2914675.

13. R. A. Memon, J. P. Li, J. Ahmed, M. I. Nazeer, M. Ismail, and K. Ali, "Cloud-Based vs. Blockchain-Based IoT: A Comparative Survey and Way Forward," *Front. Inf. Technol. Electron. Eng.*, vol. 21, no. 4, pp. 563–586, 2020, doi: 10.1631/FITEE.1800343.

14. E. Siow, T. Tiropanis, and W. Hall, "Analytics for the Internet of Things," *ACM Comput. Surv.*, vol. 51, no. 4, pp. 1–36, 2019, doi: 10.1145/3204947.

15. J. Li, Y. Zhang, Y. F. Chen, K. Nagaraja, S. Li, and D. Raychaudhuri, "A mobile phone based WSN infrastructure for IoT over future internet architecture," *Proc. - 2013 IEEE Int. Conf. Green Comput. Commun. IEEE Internet Things IEEE Cyber, Phys. Soc. Comput. GreenCom-iThings-CPSCom 2013*, pp. 426–433, 2013, doi: 10.1109/GreenCom-iThings-CPSCom.2013.89.

16. S. K. Singh, S. Rathore, and J. H. Park, "BlockIoTIntelligence: A Blockchain-enabled Intelligent IoT Architecture with Artificial Intelligence," *Futur. Gener. Comput. Syst.*, vol. 110, no. xxxx, pp. 721–743, 2020, doi: 10.1016/j.future. 2019.09.002.

17. H. Yue, L. Guo, R. Li, H. Asaeda, and Y. Fang, "DataClouds: Enabling Community-Based Data-Centric Services over the Internet of Things," *IEEE Internet Things J.*, vol. 1, no. 5, pp. 472–482, 2014, doi: 10.1109/JIOT.2014.2353629.

18. H. Rahimi, A. Zibaeenejad, and A. A. Safavi, "A Novel IoT Architecture based on 5G-IoT and Next Generation Technologies," *2018 IEEE 9th Annu. Inf. Technol. Electron. Mob. Commun. Conf. IEMCON 2018*, pp. 81–88, 2019, doi: 10.1109/IEMCON.2018.8614777.

Chapter 2

Application of Artificial Intelligence (AI) and the Internet of Things (IoT) in process industries toward Industry 4.0

Rashi Verma, Raju Yerolla, Suhailam P and Chandra Shekar Besta

National Institute of Technology, Calicut, India

CONTENTS

DOI: 10.1201/9781003407300-2

LIST OF NOTATIONS AND ABBREVIATIONS

AI   Artificial Intelligence
CPS   Cyber Physical System
ICTs   Information & Communication Technology
CPPs   Cyber Physical Production System
IoT   Internet of Things
IoS   Internet of Services
IoMs   Internet of Manufacturing Services
IoP   Internet of People

2.1 INTRODUCTION

The Industrial revolution has been deemed the most influential revolution in history due to its pervasive impact on people's everyday lives. The term "industrial revolution" has been coined for a period of history that began in eighteenth-century Great Britain and finished in the twenty-first century. As a consequence of the acceleration of technical innovation processes, this has resulted in a plethora of new tools and devices. Additionally, it includes subtler technological advancements in areas such as labor, manufacturing, and resource use. The industrialization, which commenced in the 18th century, was a turning point in human history and has continued up to the current day. We are now on the brink of a technological revolution that will impact our lives, jobs, and relationships for the rest of our lives. The transition will be unparalleled in terms of its scope, breadth, and complexity. We do not yet know how it will play out. However, one thing is certain: a thorough and coordinated response is required, which will include all stakeholders, from the public and business sectors to education and civil society. Everyone is connected to wider global developments. The industry has seen four revolutions to date [1]. Following each revolution, industry has upgraded with a new generation of technology. The chronology of the industrial revolution is shown in Figure 2.1.

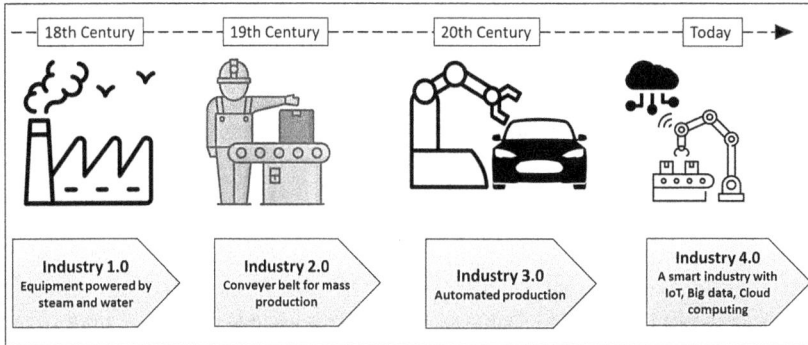

Figure 2.1 Historical perspective of industrial revolutions.

2.1.1 1st industrial revolution

The beginnings of the First Industrialization in the eighteenth century were ushered in by the discovery of steam engines and the mechanization of production. What had previously been done mechanically on a basic spinning wheel with threads now produced eight times the size in the same timeframe. The most important development in terms of increasing human productivity was the development of steam power. Where muscle power had previously been used to operate weaving looms, steam engines would be employed. This led to revolutionary breakthroughs such as the introduction of steam-powered railways, which reduce the time for individuals and items to travel over vast distances.

2.1.2 2nd industrial revolution

The Second Revolution began in the mid-nineteenth century along with the development of electricity and the introduction of production lines, the twentieth century was transformed. In one of the most significant innovations Henry Ford adapted the concept of a conveyor belt originating from the slaughterhouse to automotive manufacturing, significantly altering the process and ushering in the age of mass production. The car was now manufactured in partial phases on conveyor belts, which both accelerated the procedure and lowered the costs of production.

2.1.3 3rd industrial revolution

The rapid spread of memory-programmable controllers and computers to automate tasks in the 1970s was the driving force of the Third Industrial Revolution. Thanks to the advancement of these technologies, we can now manage the whole manufacturing process with a minimal need of human assistance. This is demonstrated by robots that follow pre-programmed orders without the need for human input.

2.1.4 4th industrial revolution

At the time of writing, the Fourth Industrial Revolution is well underway. This is referred to in the literature as "Industry 4.0" and is distinguished by the use of information and communication technology in manufacturing. These initiatives builds on the achievements of the Third Industrial Revolution Production System, which has already integrated computer technology. This is now strengthened by a network connection and has Internet-connected digital twins in certain respects. These enable systems to interact and to generate information about one another. This is the next level of industrial computerization. All systems are connected, resulting in a "cyber-physical production system" and, consequently, automated systems in which manufacturing systems, elements, and people interact over a network and output is almost self-sufficient.

Industry 4.0 has the potential to dramatically alter the manufacturing environment if all of these enablers function collectively. Examples include machines that can detect malfunctions and begin repair operations on their own, as well as self-organized logistics that can adapt to unanticipated changes in production.

Industry 4.0 is a huge success for the manufacturing sector. Manufacturing will be impacted in a variety of ways as a result of digitization, including the way in which things are created and distributed, as well as how items are managed and enhanced.

2.2 INDUSTRY 4.0

Since the first Industrial Revolution manufacturing has transitioned from water and steam power to mechanize the production machinery to electric power to create mass production. Then information technology to automat production, thereby increasing the complexity, automation, and sustainability of manufacturing processes while allowing workers to operate machinery simply, efficiently, and consistently.

In today's environment, industrial processes must be both computerized and sophisticated. In the industrial sector, mass manufacturing is increasingly being phased out in favour of personalized items. Significant advancements in manufacturing technology and applications have all contributed to increases in productivity. Industry 4.0 is characterized by a higher level of organization and supervision throughout the whole value chain of a product's life cycle, with an emphasis on more personalised customer expectations. Industry 4.0, which integrates the Internet of Things, intelligent manufacturing, and cloud-based manufacturing, remains a future, albeit feasible concept. Figure 2.2 depicts the transition of conventional industries to Industry 4.0. Industry 4.0 refers to the tight integration of humans into manufacturing processes in order to promote continuous improvement and a higher focus on value-added activities while reducing waste.

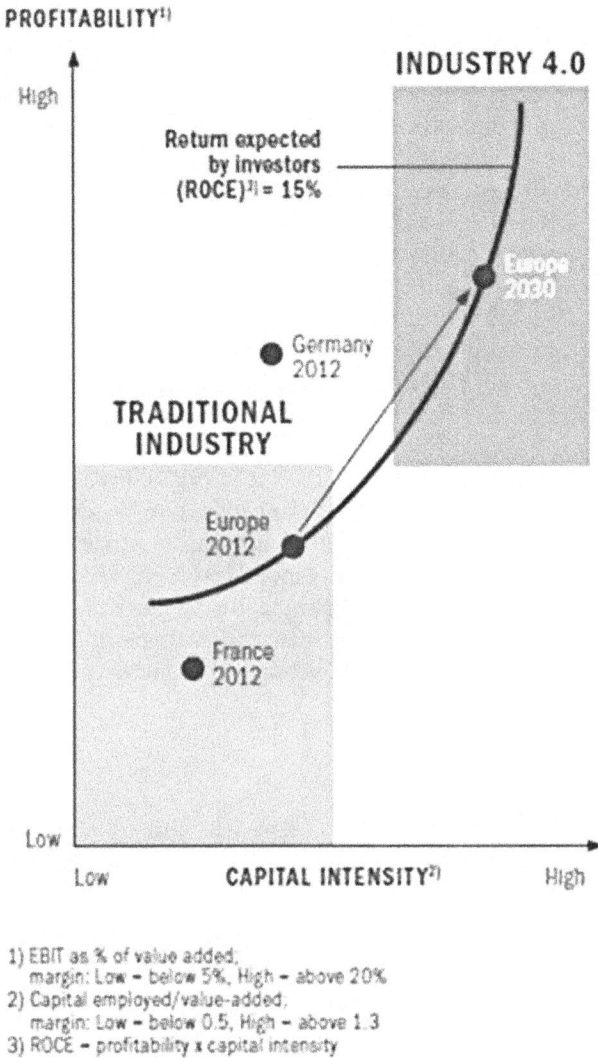

PROFITABILITY[1]

INDUSTRY 4.0

High

Return expected
by investors
(ROCE)[3] = 15%

Europe
2030

Germany
2012

TRADITIONAL
INDUSTRY

Europe
2012

France
2012

Low

Low   CAPITAL INTENSITY[2]   High

1) EBIT as % of value added;
   margin: Low – below 5%, High – above 20%
2) Capital employed/value-added;
   margin: Low – below 0.5, High – above 1.3
3) ROCE – profitability x capital intensity

Figure 2.2 Industry 4.0 will lead to higher profitability and productivity © CC
BY 3.0 [2].

2.2.1 Need for Industry 4.0

Industry 4.0's objective is to turn conventional machines into self-aware, self-learning machines capable of optimising their general efficiency and upkeep control via interaction with their environment. The purpose of Industry 4.0 is to create a platform for data-driven industrial applications that is open and cognitive. Industry 4.0's main requirements include actual data monitoring, recording the status and location of things, and retaining instructions to control industrial operations [3].

2.2.2 The Significance of Industry 4.0

Industry 4.0 principles are intended to allow businesses to work more flexibly and to analyse large volumes of data in real time, resulting in better strategic and operational decisions. This new industrial stage has been enabled by the extensive usage of ICTs in industrial settings, as well as the declining cost of electronics [4], which has resulted in the installation of more sensors in physical objects. These advances in technology have opened the path for the embedded system design and networked setup. This setup seeks to monitor and control equipment, conveyors, and products using a feedback loop that gathers a vast quantity of large amount of data and modifies the digital model by integrating data from physical processes. The outcome is a smart factor, as seen in Figure 2.3.

As a consequence, several technologies have evolved and been integrated into manufacturing systems since the 1980s. These have included cloud technology for on-demand manufacturing services, modelling for calibration, and rapid prototyping for a flexible production system. Businesses that use a Cyber Physical System (CPS) for information processing benefit from increased decision support and can respond more quickly to a range of conditions, such as breakdowns in manufacturing lines. As a consequence, through the integration of manufacturing with smart grids for energy efficiency, these systems may increase corporate productivity through the optimisation of resource use.

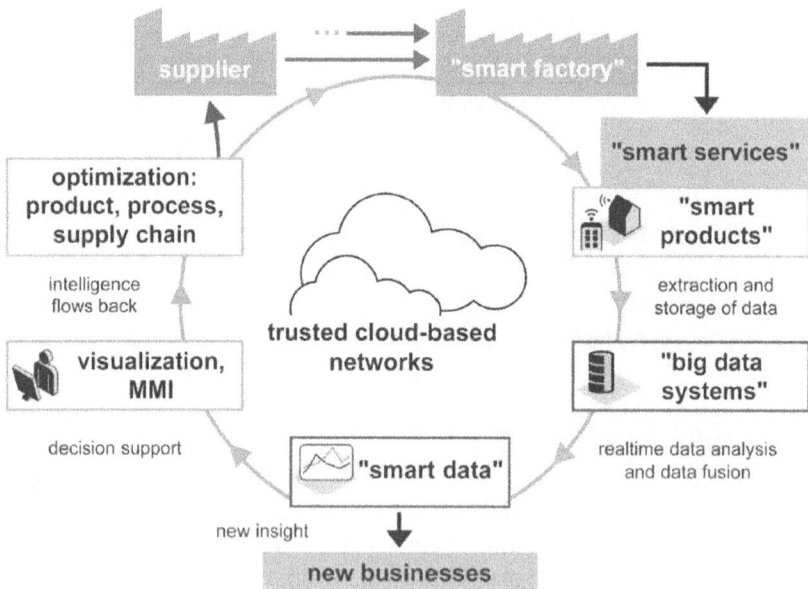

Figure 2.3 Concept of smart factories © CC BY 3.0 [2].

In addition, Industry 4.0 offers opportunities and benefits for company development. Collaborative networks of enterprises use the horizontal integration concept to pool resources, split risks, and respond swiftly to market changes, enabling them to grasp new opportunities. Figure 2.4 depicts the Industry 4.0 structure, along with its various functionalities.

2.2.3 Technologies of Industry 4.0

Industry 4.0, according to the German Federal Government, is an evolving framework where logistics and shipping networks are concerned, including structure such as the Cyber Physical Production Systems (CPPS) shown in Figure 2.5. This makes widespread use of the widely accessible information-sharing system for the automated exchange of data and in which production and business systems are consolidated [5].

Industry 4.0's nine elements, as outlined in Figure 2.7 [6], will change everything from segregated and optimised cell manufacturing into a completely automated, optimised, and integrated process. As a consequence, traditional manufacturing linkages between suppliers, producers, and consumers, as well as between people and robots, will improve and evolve [7].

2.2.4 Nine components that are bringing about the transformation to Industry 4.0

2.2.4.1 Cloud computing

The technical infrastructure that connects and communicates the sector is a cloud-based IT platform and many parts of the 4.0 Application Centres [8]. With Industry 4.0, enterprises will be required to communicate more data across locations and businesses, resulting in response times of a millisecond or less [7]. "Digital production" is a term that refers to the process of connecting several devices to a common cloud for the purpose of exchange data, and also the collection of equipment both on a production floor and throughout the whole operation (Marilungo et al. 2017).

2.2.4.2 Big data

To allow instantaneous decision-making, the gathering and detailed assessment of large amounts of data is required. A collection of data from various origins such as industrial machines and technology, as well as corporate and client solutions, will become commonplace [7]. Big data, as per Forrester, is made up of four factors: the amount of data, the variety of data, the rate at which new data is collected and processed, and data value. To find out more, collected information from previously acquired data is used find the risks that happened during different manufacturing processes in the past in

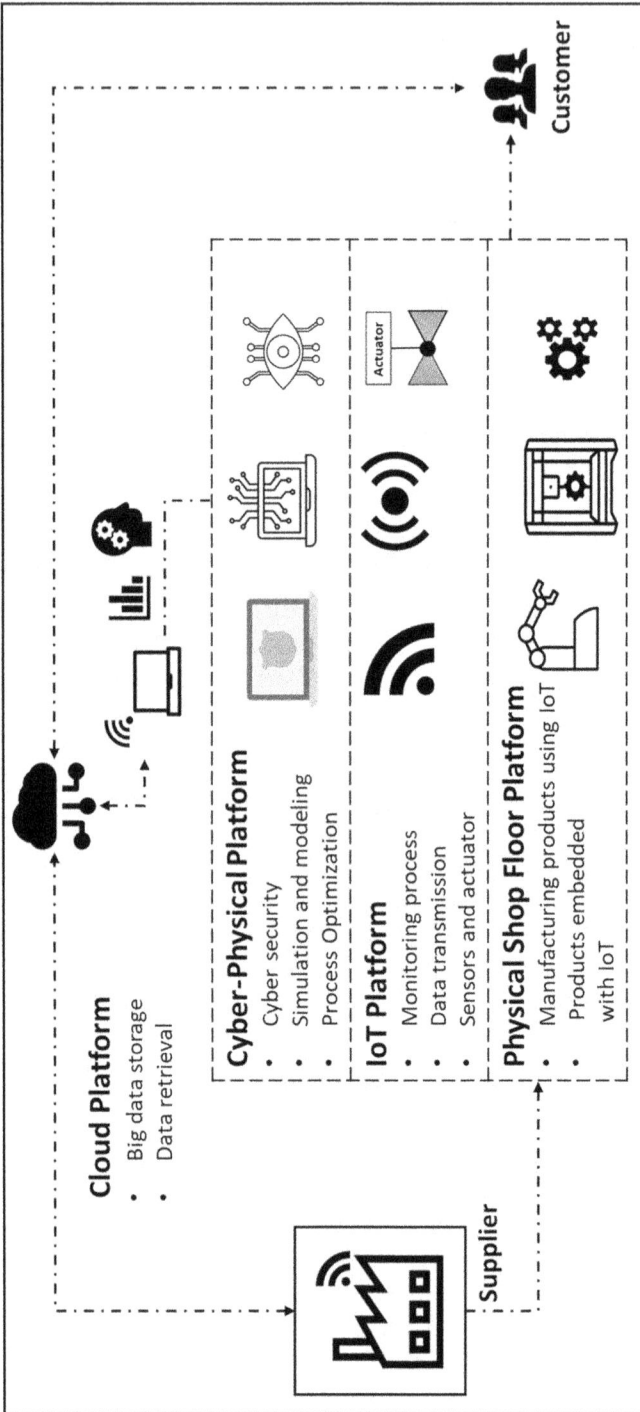

Cloud Platform
- Big data storage
- Data retrieval

Cyber-Physical Platform
- Cyber security
- Simulation and modeling
- Process Optimization

IoT Platform
- Monitoring process
- Data transmission
- Sensors and actuator

Physical Shop Floor Platform
- Manufacturing products using IoT
- Products embedded with IoT

Customer

Supplier

Figure 2.4 The Industry 4.0 paradigm.

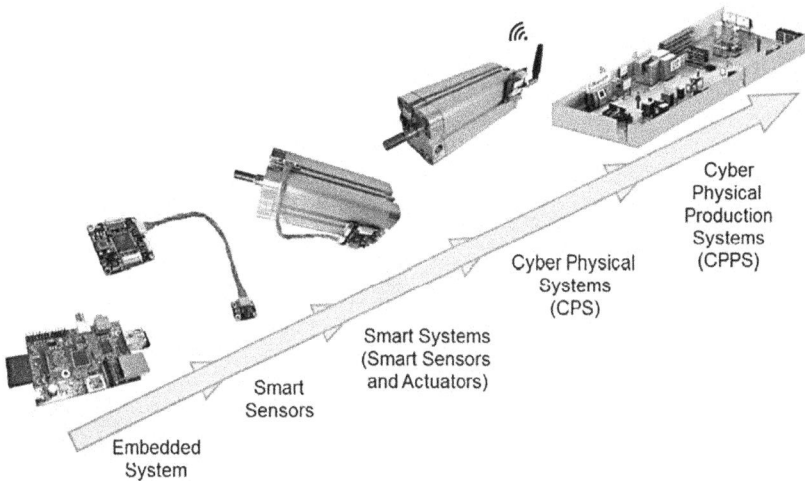

Figure 2.5 CPS – Cyber Physical Systems © CC BY 3.0 [2].

addition to anticipate challenges that may develop in the future, as well as alternative remedies to avoid them from reoccurring [10].

2.2.4.3 The Internet of Things

The Internet of Things (IoT) is a universal system of connected and evenly identified items that correspond via customary protocol [11]. The complete architecture of the IoT includes the Internet of Services (IoS), the Internet of Manufacturing Services (IoMs), the Internet of People (IoP), integrated devices, and technology and communication integration (IICT) [12]. Core parts of this component include context, limitlessness as well as effectiveness with context referring to an item's capacity to interact in a sophisticated manner with its present surroundings and respond quickly when anything changes. Optimization demonstrates the notion that today's items are greater than they used to be when simply interfaces to a human-operated system [13]. Omnipresence communicates information about an entity's location, physical or atmospheric factors, and optimizing shows that present goods are far more than just the linkages to a system of human input at the interaction point between humans and machines.

The production process ought to be smart, flexible, and linked via the mixing of materialistic goods, human determinants, inventive equipment, sensors of high intelligence manufacturing processes, and across the assembly line's organisational confines. Metrics and technology will, in future, be critical components of smart machine planning and control [12]. Integrated stacking and containers, for example, will become the motivating force underlying inventory management and storage. When it comes to freight

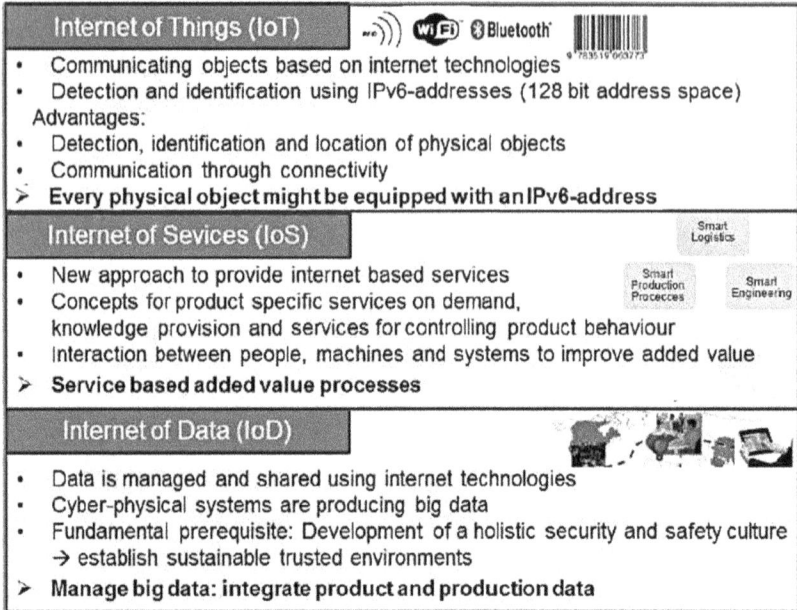

Internet of Things (IoT)

- Communicating objects based on internet technologies
- Detection and identification using IPv6-addresses (128 bit address space)

Advantages:

- Detection, identification and location of physical objects
- Communication through connectivity
- **Every physical object might be equipped with an IPv6-address**

Internet of Sevices (IoS)

- New approach to provide internet based services
- Concepts for product specific services on demand, knowledge provision and services for controlling product behaviour
- Interaction between people, machines and systems to improve added value
- **Service based added value processes**

Internet of Data (IoD)

- Data is managed and shared using internet technologies
- Cyber-physical systems are producing big data
- Fundamental prerequisite: Development of a holistic security and safety culture → establish sustainable trusted environments
- **Manage big data: integrate product and production data**

Figure 2.6 Internet of things, Internet of services, Internet of data © CC BY 3.0 [2].

transit, monitoring becomes more accurate, quicker and more secure. Figure 2.6 shows the basic elements of Industry 4.0.

2.2.4.4 Augmented reality

Industry 4.0 research is highlighting best practices for enhancing quality in assembly processes via the use of augmented reality. Augmented reality solutions might help with various tasks, such as gathering components in a warehouse and conveying maintenance procedures to mobile devices. The industry may use augmented reality to provide employees with accurate information, helping them to make better choices and carry out tasks more quickly. During the time analyzing the system in need of repair, operators may be able to obtain restoration information on how to rebuild a specific item [7].

2.2.4.5 System integration: horizontal and vertical system integration

In the field of production, the two primary strategies involved are incorporation and self-optimization [14]. Vertical integration, networked manufacturing systems, and collaborative manufacturing are the three integration aspects outlined by the Industry 4.0 paradigm: (a) horizontal integration across the whole value creation network; (b) vertical integration throughout the full network of value generation; and (c) engineering from beginning to conclusion from start to finish throughout the product life cycle.

Vertical and horizontal digital integration and computerization of industrial operation need the automation of communication and collaboration, most notably along established procedures [15].

2.2.4.6 Cybersecurity

Alongside the expanding accessibility and utilization of conventional communications protocols associated with Industry 4.0, the demand to protect crucial manufacturing technologies and production lines from cybersecurity risks has increased substantially. Consequently, secure, reliable communications are vital, as are enhanced machine and user identification and access control [7]. Closely integrating the physical, service, and digital worlds may improve the quality of information required for the design, optimization, and operation of manufacturing systems [8]. The phrase "computing, communication, and control systems" (CPS) refers to "Computing, communication, and control mechanisms are all intertwined in natural and man-made systems (physical space)" [10]. The primary features of CPS are decentralization and process autonomy.

Supply networks, which are categorized as Collaborative Cyber Physical Systems since they're commonly utilized in production system in addition to other cyber physical systems like municipal traffic control and control systems [16], are important parts of CPS development. The use of cloud technology to automatically link cyber physical systems in real time allows for continuous data flow [17]. The digital shadow of manufacturing represents a physical thing in a virtual or information realm. Massive cyber physical systems problems are used to solve the basic demand of manufacturing operations and system optimization in real settings [12]. By incorporating relevant sensors into CPS, it should be possible to identify machine failure and prepare CPS automatically for fault repair tasks. The cycle time required to conduct the task at that station is also utilized to establish the appropriate allocation of resources at each workstation [18]. Cloud computing is used in the 5C architecture to link machines (machine-to-machine or human-to-machine) [19]. For example, the smart vehicle is an example of integrated Cyber Physical System (CPS) manufacturing that represents Industry 4.0 innovation [20]. This production employs an information-mining technique to perform prognosis that attains a precision of 80% [21].

2.2.4.7 Simulation

Simulations will be increasingly widely utilized in production systems that harness real-time data to create a virtual model of the actual environment, which may include machines, goods, and workers, therefore reducing machine setup times and improving quality [7]. Virtual commissioning may be accomplished using 2D and 3D simulations, as well as models of cycle times, power consumption, and production plant layouts. During the startup phase, simulation may help decrease downtime and changes, as well as addressing production issues [21]. Additionally, simulation enhances decision-making by simplifying and expediting it [14].

2.2.4.8 Autonomous robot

On a daily basis, robots become increasingly self-sufficient, flexible, and coordinated, and they will unavoidably connect with one another and operate safely among others while learning from experience [7]. An autonomous robot is used to improve the accuracy of autonomous manufacturing processes and to operate in locations where there are few people. Autonomous robots are capable of performing jobs accurately and rationally in a certain amount of time [22] while also prioritizing safety, adaptability, diversity, and participation, Table 2.1 shows some of the examples.

2.2.4.9 Additive manufacturing

Additive manufacturing is primarily used in Industry 4.0 to create more personalized items with complex and lightweight constructions. If additive manufacturing is decentralized, the cost of production will decrease, resulting in a greater performance rate [7]. Production should be quicker and less costly through the use of different manufacturing methods such as fused deposition modeling (FDM), selective laser melting (SLM), and selective laser sintering (SLS) [8]. Numerous businesses confront the challenge of expanding product customization and decreasing time to market as customer demands evolve on a regular basis. Digitization, IT sector involvement, and correct processes were able to resolve the aforementioned issues. A more efficient organizational structure was required to accommodate growing demand while also extending the life of the product.

For instance, identical model automobiles are offered with a variety of engine, bodywork, and equipment choices, all designed to satisfy the demand of more savvy clients who are selective [13]. Figure 2.7 depicts the various technologies of Industry 4.0.

Table 2.1 Example of autonomous robots

Serial No.	Robot name	Manufacturing company	Functioning
1	Kuka LBR iiwa	Kuka	A lightweight robot is designed for critical industrial tasks
2	Baxter	Rethink Robotics	Packaging robot with interactive capabilities
3	BioRob Arm	Bionic robotics	Used in areas close to human

Simulation

Robot

Big Data
&
IoT

Additive
Manufacturing

Industry 4.0
Technologies

Cloud
Computing

Cyber
security

System
Integration

Augmented
Reality

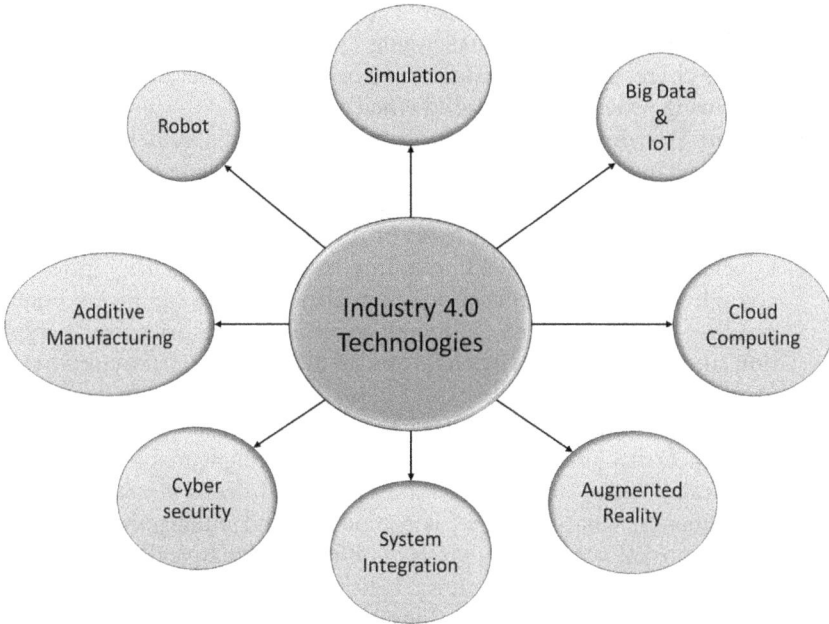

Figure 2.7 Industry 4.0 technologies.

2.2.5 Issues and challenges faced by Industry 4.0

Mechanical systems have been used since the dawn of time, and today's highly automated assembly lines are, to some extent, a continuation of this process. New technologies have benefited industry development by enabling it to be flexible and adaptive to shifting market needs. Embed ability, predictability, adaptability, and robustness are just a few of the unforeseen conditions that may arise. Numerous complications and basic concerns develop with the program's implementation of "Industry 4.0" [12].

Intelligent Decision-Making and Negotiation Mechanisms: The two most critical characteristics of a smart factory are autonomy and sociability. However, since the majority of systems adhere to the fundamental 3C paradigm (Company-Client-Competitor), they lack these two features. Further research is necessary to develop an autonomous manufacturing architecture in lieu of the hierarchical structure [12].

- High-Speed IWN Protocols: Industry needs a robust and reliable communication network capable of transporting large amounts of data. However, the IWN (Industrial Wireless Protocols) standards WIP-PA and Wireless HART do not offer sufficient bandwidth, making them inefficient and slow [12].
- Production-specific Big Data and Analytics: While every industry generates a significant quantity of data, assuring the quality and integrity

of data that focuses on the usable characteristics of data captured by production systems is a challenging challenge. By including just the unique characteristics of data, it may be possible to ascertain the manufacturing process's quality and efficiency.

- System Modeling and Analysis: Any system may be modeled as a self-organized manufacturing process using the appropriate dynamic equation and control model. However, this is only viable for basic systems; larger systems remain a hot focus of research [12].
- Cybersecurity: Industry 4.0 requires robust cybersecurity prior to going live. Any kind of cyber attack might result in a significant financial loss for the business. Although some smart factories use encryption and authorization mechanisms as a form of cybersecurity, this may prove insufficient [7].
- Modularize and Flexibility Physical Artifacts: In order to manufacture a product, machining and testing equipment ought to be shared and cooperated on in order to facilitate dispersed decision-making. As a consequence, what is needed is a modular and intelligent conveying device capable of changing production routes dynamically [12].
- Investment Issues: Investment is a relatively broad problem that affects the majority of new technology-based manufacturing endeavors. Significant expenditure is necessary to adopt industry standards. 4.0 are originally a SME (Structural Equation Modelling). Implementing all of Industry 4.0's pillars requires a significant investment from an industry [23].

2.3 ARTIFICIAL INTELLIGENCE

Artificial Intelligence (AI) is a vast subject of computer science focused on the development of intelligent computers capable of undertaking tasks that would typically need human intelligence. While Artificial Intelligence (AI) has been defined in a variety of ways over the past few decades, John McCarthy recommends the following definition in this 2004 study: "It is the science and engineering of creating intelligent devices, particularly intelligent computer programs." This is comparable to the analogous challenge of utilizing computers to comprehend human intellect, although AI is not limited to physiologically observable techniques. The Artificial Intelligence issue was founded, however, by Alan Turing's landmark 1950 paper "Computing Machinery and Intelligence." In this paper, Turing, who has been called the "Father of Computer Science," answers the topic "Can Machines Think?" Artificial Intelligence was the answer to this very question. Artificial Intelligence is an area of computer science whose purpose is to answer Turing's question in the affirmative. Researchers working on Artificial Intelligence are attempting to mimic or emulate human intellect in robots.

2.3.1 Artificial Intelligence: a brief history

Smart robots and artificial beings made their first appearance in Greek mythology. Aristotle's invention of the syllogism and its application of rational thinking marked a turning point in civilization's quest to understand its own mind. Despite its lengthy and illustrious history, Artificial Intelligence as we comprehend it now has only been around for about a century. Here's a rundown of some of the most significant AI events in recent years.

1940s
- "A Logical Calculus of Concepts Intrinsic in Nervous Action," by Warren McCullough and Walter Pitts, was published in 1943. This was the first mathematical formalism for building a learning algorithm was described in this study.
- *The Organization of Behavior: A Neuropsychological Theory*, written by Donald Hebb in 1949, proposed that experiences alter brain circuits and that connections between neurons become stronger as they are utilized more frequently.

1950s
- In 1950, Alan Turing publishes "Computing Machinery and Intelligence," in which he proposes the Turing Test, a technique for determining whether or not a computer is intelligent.
- In 1950, two Harvard freshmen, Marvin Minsky and Dean Edmonds, build SNARC (Stochastic Neural Analog Reinforcement Calculator), the first neural network computer.
- In 1950, Claude Shannon releases "Programming a Computer for Chess Playing."
- Isaac Asimov wrote "The Three Laws of Robotics" in 1950.

1960–2010
- In 1963, John McCarthy founds the Stanford Artificial Intelligence Laboratory.
- In 1972, PROLOG, a logic programming language, is created.
- Japan's Ministry of International Trade and Industry (MITI) launched the Fifth Generation Computer Systems project in 1982.
- In response to Japan's FGCS, the US government launched the Strategic Computing Initiative in 1983, which funds advanced computing and Artificial Intelligence research via DARPA. DART, an automated logistical planning and scheduling program, was used by US troops during the 1991 Gulf War.
- In 1997, Gary Kasparov, the world chess champion, is beaten by IBM's Deep Blue.
- In 2005, STANLEY, a self-driving car, wins the DARPA Grand Challenge.

- In 2008, Google advances voice recognition technology and makes it accessible through its iPhone app.

2010–2020
- Apple introduced Siri, an Artificial Intelligence-powered virtual assistant, as part of their iOS operating system in 2011.
- In 2014, Google's self-driving vehicle passes the state's driving test for the first time.
- Amazon launched Alexa, a virtual assistant, in 2014.
- In 2016, Hanson Robotics introduces Sophia, the world's first "robot citizen," a humanoid robot capable of face recognition, linguistic communication, and facial expression.
- Baidu will make its Linear Fold AI algorithm accessible to scientists and physicians working on a SARS-CoV-2 vaccine during the early phases of the pandemic in 2020. In only 27 seconds, the algorithm can predict the virus's RNA sequence, which is 120 times quicker than previous approaches.

While Artificial Intelligence is a diverse discipline with a variety of methodologies, advances in algorithms are causing a paradigm shift in practically every aspect of the IT sector. AI systems now have a wide range of real-world applications.

2.3.2 Approach of Artificial Intelligence in Industry 4.0

Artificial Intelligence (AI) is a branch of machine learning that involves image analysis, text classification, automation, and advanced analytics. Broadly speaking, machine learning and Artificial Intelligence have been seen as "black-art" methods, with a dearth of convincing evidence to persuade business that these techniques would work regularly in a consistent manner while offering a financial return on investment simultaneously, the efficiency of learning algorithms is extremely reliant on the programmer expertise and choices. As a consequence, the success of Artificial Intelligence in industrial applications is restricted. Industrial AI, on the other hand, is a field devoted to developing, verifying, and implementing varied algorithms are developed for industrial applications with lengthy productivity.

It's a useful method and discipline for producing answers for industrial implementation, providing a link between academic and industry AI development. Automation powered by Artificial Intelligence has a substantial quantitative influence on the rise of productivity. In addition, today's industries face new market demands and competitive pressures.

They need a paradigm shift dubbed Industry 4.0. Integrating AI with new technologies such as the Industrial Internet of Things (IIoT) [24], Big Data analytics [20, 24–26], cloud computing [27, 28], and cyber physical systems [10, 10, 12] will enable businesses to function more efficiently, effectively, and sustainably.

Due to the fact that industrial AI is still in its infancy, its structure, techniques, and difficulties must be defined in order to provide a foundation for its use in industry. To do this, we developed an Industrial AI ecosystem that incorporates the field's most critical components and acts as a road map for increased knowledge and application. The methods and technologies on which an Industrial AI system may be constructed are also discussed.

2.3.3 Key elements in Industrial AI

ABCDE: The acronym 'ABCDE' may be used to define the critical components of Industrial AI. The important variables are analytics technology (A), Big Data technology (B), cloud or cyber technology (C), domain expertise (D), and evidence (E) [29]. While research is at the basis of AI, it is only beneficial when combined with other components. Both Big Data technology and cloud computing are key components that serve as both a data source and a foundation for Artificial Intelligence in industry. While they are key components, domain expertise and proof are also critical concerns that are sometimes ignored in this scenario. Domain expertise is critical in the following areas:

1. Recognizing the issue and directing Industrial AI's capabilities toward resolving it;
2. Comprehending the system in order to acquire data of the appropriate quality;
3. Understanding the technical meanings of the metrics and how they correlate to the fundamental properties of a system or processes;
4. Recognizing how these factors vary amongst systems.

Evidence is also required to validate Industrial AI models and to include learning through time capabilities into them. We can only develop the AI model over time by amassing trends in data and the information (or label) that goes with them. Figure 2.8 illustrates how AI may take us from visible to invisible space, and from resolving difficulties to preventing them from occurring in the first place.

2.3.4 Industrial AI ecosystem

The proposed Industrial AI ecosystem, as depicted in Figure 2.9, is a progressive approach for thinking for determining industry demands, issues, technologies, and strategies in order to build revolutionary AI systems. This diagram may be used as a logical framework for developing and implementing an Industrial AI strategy. This ecosystem highlights common unmet business demands such as self-awareness, self-comparison, self-prediction, self-optimization, and resilience.

When seen through the lens of the Cyber Physical Systems (CPS) proposed in [2], these four technologies become more intelligible. The above mentioned

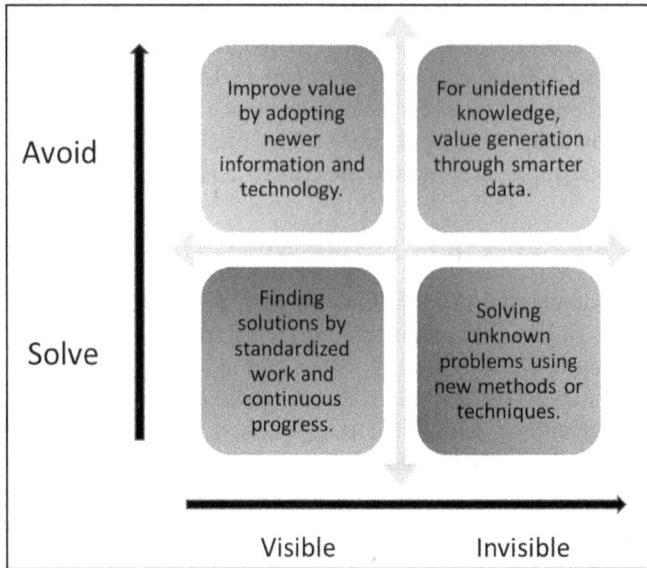

Figure 2.8 Industrial AI's impact: from resolving apparent issues to avoiding problems those aren't evident.

four in the 5C era, technology is the key to success. (Connection, Conversion, Cyber, Cognition, and Configuration), as shown in Figure 2.10.

2.3.5 Limitations of industrial Artificial Intelligence

Industrial AI is accompanied with a plethora of expectations; even if only some of these expectations are met, it will result in a "one-of-a-kind" experience and substantial difficulties in deploying AI to industries. [29] Among the present obstacles and complications, the following concerns and complexities are more important and sorted:

1. Interactions between machines: While AI algorithms can properly convert a set of inputs to a set of outputs, they are vulnerable to minute input discrepancies caused by machine variances. It must guarantee that future implementations of specific Artificial Intelligence (AI) solutions do not interfere with the operation of other systems.
2. Data integrity is one of the most important factors to consider. Large, clean data sets with low errors are required for AI systems. The downstream repercussions of learning from erroneous or insufficient data sources may be contaminant
3. Information security: The smart manufacturing system is becoming progressively prone to cyber attack as more linked technologies are used. The scope of this vulnerability is currently underestimated, and the industry is unprepared to deal with the security threats that ensue [11].

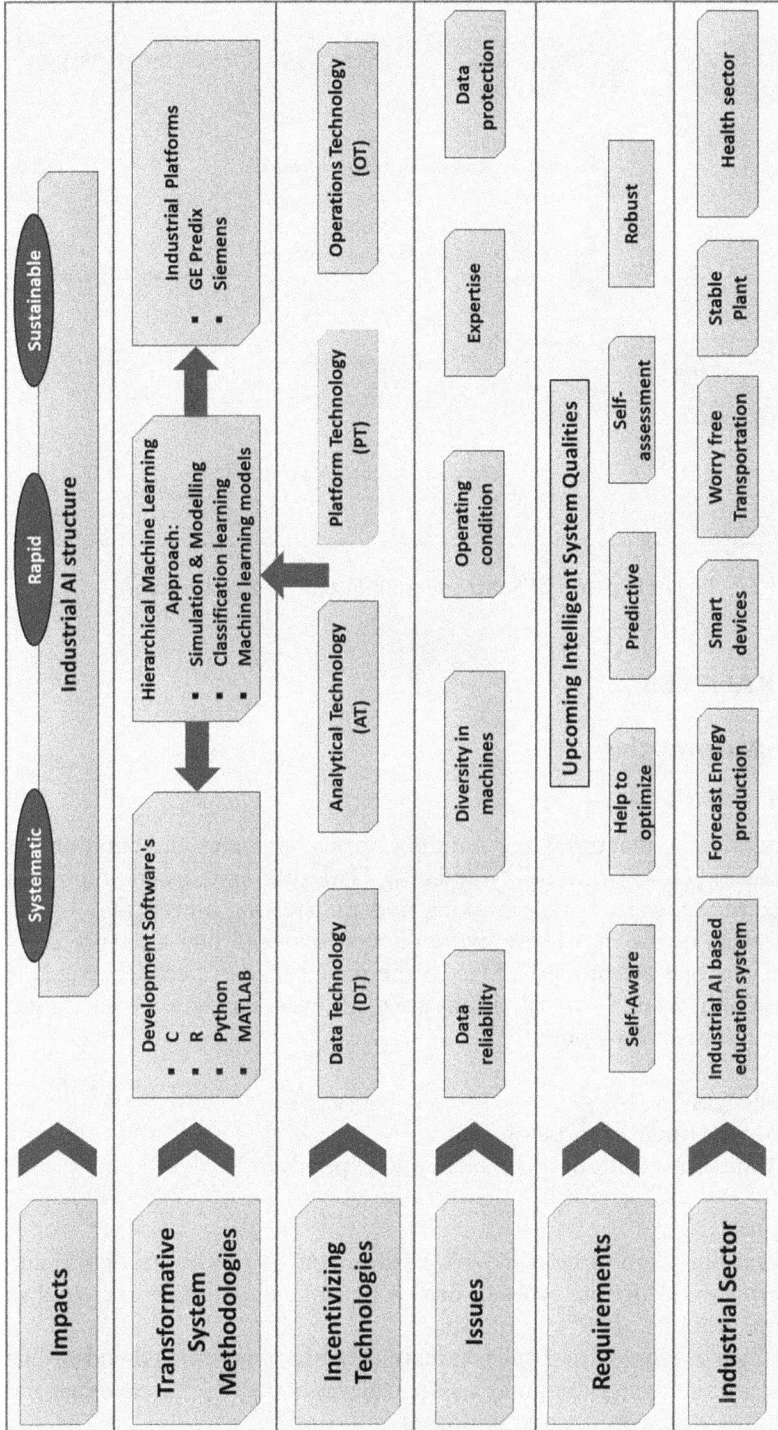

Figure 2.9 Industrial AI eco-system.

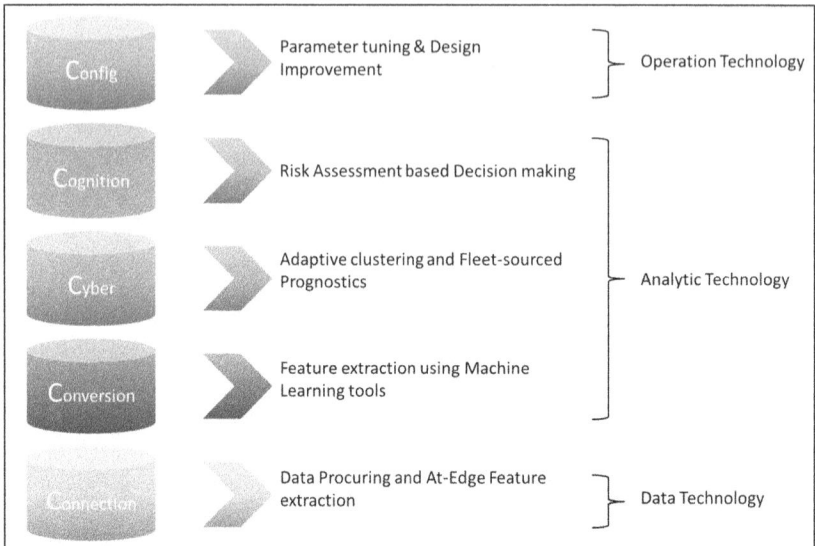

Figure 2.10 Technology for CPS introduction in the industrial sector.

2.4 EXAMPLES

2.4.1 Automotive

2.4.1.1 Description

Automobile manufacturers use a number of technologies and components to enhance their vehicles' utility and cost. Hydraulic systems are important because they govern vehicle braking and the steering wheel, two significant aspects of driving. These hydraulic systems need high-pressure pipes or hoses to operate properly. Due to the wide range of vehicle designs, a plethora of hydraulic hose variations are made every day to serve busy automotive manufacturing lines.

Challenges
• A wide range of those options.
• Manual inspection is inefficient and expensive.

Solution
• Computer vision models have been taught to reliably recognise and categorise different types of defects (cuts, gouges, drag marks, and so on) at a rate of 99% or higher.
• When compared to a fully human method, an automated end-of-line inspection using computer vision achieves almost 100% coverage and improves categorization uniformity.

2.4.2 Food and beverage

2.4.2.1 Description

While making bread or pastries at home may seem to be a simple process capable of being handled by the majority of people, producing baked goods on a large scale offers a number of challenges. The aesthetic of the baked product is essential in the client's eyes, and many returns are based on how a product appears rather than how it tastes.

Challenges
- One-on-one examination by a worker is prohibitively expensive.
- Escaping faults is considerably more expensive than non-escaping defects due to increased expenses in packing, shipping, and inventory management.

Solution
- The rapid training of computer vision models for the purpose of identifying damaged items.
- Side cameras with an edge device and stream processing provide for real-time computing and 100% inspection at line speeds.

2.4.3 Medical equipment

2.4.3.1 Description

Pharmaceutical products are complicated and are frequently mass-produced on automated lines capable of outputting 100 million or more units per year. To increase profitability, several pharmaceutical businesses are adopting digital transformation. The use of cameras and streaming analytics tools to help with fault identification on high-throughput lines is one aspect of this trend.

Challenges
- Manual inspection is time-consuming and ineffective in detecting some sorts of problems.
- High rejection product cost.
- Limited line availability and throughput.

Solution
- Computer vision enables the detection of problems that are not readily apparent to the naked eye.
- 99% accuracy in identifying and classifying defects.

2.5 CONCLUSION

Quicker computers, smarter machines, smaller sensors, and less expensive data storage and transmission may allow machines and other things to interact with and adapt to one another. The nine components of Industry 4.0 are described in depth, along with examples to help understand their relevance and identify the obstacles and issues associated with their implementation. As Industry 4.0 adoption grows, additional research streams, such as accessible and manufacturing administration and a well-structured distribution network, information gathering from manufacturing lines, and optimization of that input for an application in appropriate equipment, energy efficiency, and strategic planning, should be produced.

Industry 5.0 is a term that refers to research disciplines considered part of the next industrial age. However, it also refers to a broader transition that includes effects on social order, government and institutions, human personalization, and economic and industrial effects. This section explores the industrial applications of Artificial Intelligence and the Internet of Things. Since Artificial Intelligence and the Internet of Things have advanced from science fiction to the forefront of game-changing innovations, there is an urgent need for systematic Artificial Intelligence and the Internet of Things research and implementation to track their true impact on Industry 4.0, which refers to the next wave of technological systems. This chapter explains how AI and IoT fit into the Industry 4.0 concept. By providing an insight into the Industrial AI and IoT ecosystem in current production, this chapter also intends to give a foundation for conceptualizing steps toward adopting Industrial AI and IoT technologies.

ACKNOWLEDGEMENTS

We want to thank our faculty-in-charge of the National Institute of Technology, Calicut, for guiding us in writing this book chapter. We would also like to thank the reviewer for doing a detailed review and giving useful comments.

REFERENCES

1. E. Oztemel and S. Gursev, "Literature review of Industry 4.0 and related technologies," *J. Intell. Manuf.*, vol. 31, no. 1, pp. 127–182, Jan. 2020, doi: 10.1007/s10845-018-1433-8.
2. M. Crnjac, I. Veža, and N. Banduka, "From Concept to the Introduction of Industry 4.0," *Int. J. Ind. Eng. Manag.*, vol. 8, no. 1, pp. 21–30, 2017.
3. F. Almada-Lobo, "The Industry 4.0 revolution and the future of Manufacturing Execution Systems (MES)," *J. Innov. Manag.*, vol. 3, no. 4, Art. no. 4, 2015, doi: 10.24840/2183-0606_003.004_0003.
4. A. Schumacher, S. Erol, and W. Sihn, "A Maturity Model for Assessing Industry 4.0 Readiness and Maturity of Manufacturing Enterprises," *Procedia CIRP*, vol. 52, pp. 161–166, Jan. 2016, doi: 10.1016/j.procir.2016.07.040.

5. M. A. K. Bahrin, M. F. Othman, N. H. N. Azli, and M. F. Talib, "Industry 4.0: A Review On Industrial Automation AND Robotic," *J. Teknol.*, vol. 78, no. 6–13, Art. no. 6–13, Jun. 2016, doi: 10.11113/jt.v78.9285.
6. S. Parhi, K. Joshi, T. Wuest, and M. Akarte, "Factors Affecting Industry 4.0 Adoption – A Hybrid SEM-ANN Approach," *Comput. Ind. Eng.*, vol. 168, p. 108062, Jun. 2022, doi: 10.1016/j.cie.2022.108062.
7. M. Rüßmann, M. Lorenz, P. Gerbert, M. Waldner, J. Justus, and M. Harnisch, "Industry 4.0: The Future of Productivity and Growth in Manufacturing Industries," vol. 9, p. 14, 2015.
8. M. Landherr, U. Schneider, and T. Bauernhansl, "The Application Center Industrie 4.0 - Industry-driven Manufacturing, Research and Development," *Procedia CIRP*, vol. 57, pp. 26–31, Jan. 2016, doi: 10.1016/j.procir.2016.11.006.
9. E. Marilungo, A. Papetti, M. Germani, and M. Peruzzini, "From PSS to CPS Design: A Real Industrial Use Case Toward Industry 4.0," *Procedia CIRP*, vol. 64, pp. 357–362, Jan. 2017, doi: 10.1016/j.procir.2017.03.007.
10. B. Bagheri, S. Yang, H.-A. Kao, and J. Lee, "Cyber-Physical Systems Architecture for Self-Aware Machines in Industry 4.0 Environment," *IFAC-Pap.*, vol. 48, no. 3, pp. 1622–1627, Jan. 2015, doi: 10.1016/j.ifacol.2015.06.318.
11. P. Leitão, S. Karnouskos, L. Ribeiro, J. Lee, T. Strasser, and A. W. Colombo, "Smart Agents in Industrial Cyber–Physical Systems," *Proc. IEEE*, vol. 104, no. 5, pp. 1086–1101, May. 2016, doi: 10.1109/JPROC.2016.2521931.
12. J. Qin, Y. Liu, and R. Grosvenor, "A Categorical Framework of Manufacturing for Industry 4.0 and Beyond," *Procedia CIRP*, vol. 52, pp. 173–178, Jan. 2016, doi: 10.1016/j.procir.2016.08.005.
13. K. Witkowski, "Internet of Things, Big Data, Industry 4.0 – Innovative Solutions in Logistics and Supply Chains Management," *Procedia Eng.*, vol. 182, pp. 763–769, Jan. 2017, doi: 10.1016/j.proeng.2017.03.197.
14. G. Schuh, T. Potente, C. Wesch-Potente, A. R. Weber, and J.-P. Prote, "Collaboration Mechanisms to Increase Productivity in the Context of Industrie 4.0," *Procedia CIRP*, vol. 19, pp. 51–56, Jan. 2014, doi: 10.1016/j.procir.2014.05.016.
15. S. Erol, A. Jäger, P. Hold, K. Ott, and W. Sihn, "Tangible Industry 4.0: A Scenario-Based Approach to Learning for the Future of Production," *Procedia CIRP*, vol. 54, pp. 13–18, Jan. 2016, doi: 10.1016/j.procir.2016.03.162.
16. D. Ivanov, B. Sokolov, and M. Ivanova, "Schedule Coordination in Cyber-Physical Supply Networks Industry 4.0," *IFAC-Pap.*, vol. 49, no. 12, pp. 839–844, Jan. 2016, doi: 10.1016/j.ifacol.2016.07.879.
17. T. Stock and G. Seliger, "Opportunities of Sustainable Manufacturing in Industry 4.0," *Procedia CIRP*, vol. 40, pp. 536–541, Jan. 2016, doi: 10.1016/j.procir.2016.01.129.
18. D. Kolberg and D. Zühlke, "Lean Automation Enabled by Industry 4.0 Technologies," *IFAC-Pap.*, vol. 48, no. 3, pp. 1870–1875, Jan. 2015, doi: 10.1016/j.ifacol.2015.06.359.
19. D. S. de Dutra and J. R. Silva, "Product-Service Architecture (PSA): Toward a Service Engineering Perspective in Industry 4.0," *IFAC-Pap.*, vol. 49, no. 31, pp. 91–96, Jan. 2016, doi: 10.1016/j.ifacol.2016.12.167.
20. J. Shi, J. Wan, H. Yan, and H. Suo, "A survey of Cyber-Physical Systems," in *2011 International Conference on Wireless Communications and Signal Processing (WCSP)*, Nov. 2011, pp. 1–6. doi: 10.1109/WCSP.2011.6096958.

21. S. Simons, P. Abé, and S. Neser, "Learning in the AutFab – The Fully Automated Industrie 4.0 Learning Factory of the University of Applied Sciences Darmstadt," *Procedia Manuf.*, vol. 9, pp. 81–88, Jan. 2017, doi: 10.1016/j.promfg.2017.04.023.

22. S. Vaidya, P. Ambad, and S. Bhosle, "Industry 4.0 – A Glimpse," *Procedia Manuf.*, vol. 20, pp. 233–238, Jan. 2018, doi: 10.1016/j.promfg.2018.02.034.

23. T.-T. Shi, P. Li, S.-J. Chen, Y.-F. Chen, X.-W. Guo, and D.-G. Xiao, "Reduced Production of Diacetyl by Overexpressing BDH2 Gene and ILV5 Gene in Yeast of the Lager Brewers with One ILV2 Allelic Gene Deleted," *J. Ind. Microbiol. Biotechnol.*, vol. 44, no. 3, pp. 397–405, 2017, doi: 10.1007/s10295-017-1903-6.

24. L. D. Xu, W. He, and S. Li, "Internet of Things in Industries: A Survey," *IEEE Trans. Ind. Inform.*, vol. 10, no. 4, pp. 2233–2243, Nov. 2014, doi: 10.1109/TII.2014.2300753.

25. J. Lee, E. Lapira, B. Bagheri, and H. Kao, "Recent Advances and Trends in Predictive Manufacturing Systems in Big Data Environment," *Manuf. Lett.*, vol. 1, no. 1, pp. 38–41, Oct. 2013, doi: 10.1016/j.mfglet.2013.09.005.

26. J. Lee, H. D. Ardakani, S. Yang, and B. Bagheri, "Industrial Big Data Analytics and Cyber-physical Systems for Future Maintenance & Service Innovation," *Procedia CIRP*, vol. 38, pp. 3–7, Jan. 2015, doi: 10.1016/j.procir.2015.08.026.

27. D. Wu, M. J. Greer, D. W. Rosen, and D. Schaefer, "Cloud Manufacturing: Strategic Vision and State-of-the-Art," *J. Manuf. Syst.*, vol. 32, no. 4, pp. 564–579, Oct. 2013, doi: 10.1016/j.jmsy.2013.04.008.

28. L. Zhang et al., "Cloud Manufacturing: A New Manufacturing Paradigm," *Enterp. Inf. Syst.*, vol. 8, no. 2, pp. 167–187, Mar. 2014, doi: 10.1080/17517575.2012.683812.

29. J. Lee, H. Davari, J. Singh, and V. Pandhare, "Industrial Artificial Intelligence for Industry 4.0-Based Manufacturing Systems," *Manuf. Lett.*, vol. 18, pp. 20–23, Oct. 2018, doi: 10.1016/j.mfglet.2018.09.002.

Chapter 3

A review on edge computing

Working, comparisons, benefits, vision, instances and illustrations along with challenges

Parul Dubey

G H Raisoni College of Engineering, Nagpur, India

Pushkar Dubey and Kailash Kumar Sahu

Pandit Sundar Lal Sharma (Open) University, Bilaspur, Chhattisgarh, India

CONTENTS

3.1 INTRODUCTION

Edge computing, also known as distributed information technology architecture (ITA), is an unique advantage in which client data is processed close to the point where it was received by the server.

Currently, research is an important aspect of every organization, providing critical business information and allowing the real-time administration of critical company operations and activities. built-in sensors and Internet of Things (IoT) devices that function in real time from remote locations or in difficult working circumstances have the potential to acquire massive amounts of data from today's businesses on a daily basis, allowing them to become more efficient.

DOI: 10.1201/9781003407300-3

As a result of these developments, businesses are reevaluating their approach to computer use and management. Conventional computing paradigms focused on central data centers and the internet as we know it now are incapable of coping adequately with the massive volumes of data to be found in the real world. These efforts may be thwarted by network constraints such as bandwidth restrictions, latency restrictions, and unpredictability restrictions. Enterprises are benefiting from the introduction of edge computing architecture to achieve address these data challenges.

Essentially, edge computing is the practice of storing information closer to the place of origination rather than in a centralized data center. It is found to be more efficient to process and analyze data locally as it is produced on the ground than to having it transmitted to a centralized data centre for processing and analysis. An example: As soon as edge computing is done, the results, which include real-time business insights and equipment maintenance estimates, are routed back to the main data centre for review and human involvement. It should come as no surprise that edge computing is gaining attention since it is transforming IT and business computing.

In order to process data as near to an Internet of Things device as feasible, edge computing is utilized. As a consequence, time, efficiency, budget, and security may all be enhanced for organizations' IT departments. There are a number of other essential technologies that operate in conjunction with edge computing, including hybrid cloud and 5G. Devices and applications that are part of the Internet of Things (IoT) obtain significant benefits from this technology. In truth, the Internet of Things and the edge are more than just good combinations; they are becoming increasingly intertwined.

3.2 LITERATURE REVIEW

In the wake of the rise of the Internet of Things (IoT) and the popularity of cloud-based services, a new computing paradigm, known as edge computing, has emerged. As even the Internet of Things (IoT) and cloud services grow more widespread, edge computing is becoming increasingly popular. Some of the challenges that edge computing may address include security and privacy, reaction speed and battery life to mention a few [1] Numerous examples of edge computing in operation are provided throughout the chapter, spanning beyond cloud dumping to smart urban and collaborative edge computation. Edge computing difficulties and potential are explored in this chapter, which experts believe will catch the attention of the community and stimulate more study in this area.

Mobile edge computing (MEC) has the potential to offload Internet of Vehicles (IoV) applications while also allocating resources. The majority of current offloading methods are useless because they consider the application as a whole when determining whether or not it should be offloaded. A more sophisticated solution is presented in the paper [2], which is an application-centric architecture with a finer-grained offloading method based on the separation of apps. Nodes in the framework represent subtasks, and edges between nodes indicate data flow relationships between subtasks; this is how we build applications in our framework. It is possible to find offloading nodes in both automobiles and MEC servers that are within range of the network. As part of the offloading process, subtasks are allocated to various computing nodes, and the procedure is repeated. Furthermore, offloading takes care of the time constraints connected with each subtask. It has been shown via experimentation that the proposed offloading approach outperforms alternative non-partitioning methods in terms of both execution time and throughput.

For consumers who want to access data quickly, multi-access edge computing (MEC) is an additional storage option. It is not possible for individual edge servers to store huge volumes of data due to their physical dimension and storage capacity limitations. As a result of the nature of their operation, the edge servers often have to communicate data to other servers in order to deliver services to end users. They operate in a somewhat given environment and found that they are administered by a range of edge network operators. One of the most significant obstacles to collaborative edge storage is the lack of incentives and trust among the participants [3]. These difficulties are addressed in this work by the proposal of a new decentralised system, CSEdge, which uses blockchain technology to facilitate collaborative edge storage. To encourage competition among CSEdge servers, edge servers may submit data offloading requests to other CSEdge servers. Prior successes and goodwill are taken into consideration when selecting the prize winners. Their ability to store offloaded data will be recognised, and they'll be rewarded for the successful completion of a difficult task. Their reputation will be assessed in the future on the basis of their achievements, which will be recorded on the blockchain via a distributed consensus process. This infrastructure is built on Hyperledger Sawtooth and has been tested in a simulated MEC against a baseline system as well as against two current state-of-the-art systems. Based on the results, CSEdge is capable of facilitating edge server cooperation on storage in a cost-effective and efficient manner.

Because of the prevalence of the Internet of Things (IoT) and cloud services, edge computing allows data processing to take place at the edge of the network rather than solely in the cloud [4]. Edge computing may be able to handle issues such as latency, power consumption on mobile devices, bandwidth charges, security, and privacy, along with many other things.

One developing architecture, Mobile Edge Computing (MEC), extends cloud computing capabilities to the network edge via the use of mobile base stations [5]. It may be used in cellular, wireless, and wired applications since it makes use of software based at the network's edge. Mobile subscribers, enterprises, and other vertical industries all benefit from MEC's flawless integration of multiple application service providers and suppliers, which is available to all customers. It is a vital component of the 5G architecture, since it allows for the implementation of a broad variety of cutting-edge applications and services that need very rapid response times. The paper covered recent research and technological advancements in the area of microelectromechanical systems. There is an explanation of what MEC is, what its advantages are, how it may be implemented, and some of the probable applications.

Because of the extraordinary advancements in embedded systems-on-a-chip technology, many commercial devices now have sufficient processing power to run full operating systems. This has led to an increase in the potential of the Internet of Things. Early Internet of Things (IoT) devices were only capable of gathering and transmitting data for the purpose of analysis. Because of the increasing processing capacity of current devices, however, it is now feasible to do computations on-site, a process which is referred to as edge computing [6]. Edge computing, which allows cloud computing to provide a greater variety of services and applications by bringing services in response to the network's edge, is expanding its capabilities. Edge computing technologies are currently the subject of our research; as part of this inquiry, researchers highlight and report on recent breakthroughs in this field. Because of this classification and categorization, we have been able to determine the most significant and supporting characteristics of several edge computing paradigms for the Internet of Things.

3.3 THE WORKING OF EDGE COMPUTING

It is the only part of edge computing that takes into account the physical location of the device. In traditional corporate computing, data is usually made at the client port, such as a person's computer, before it is sent out to other people. In this case, the data is distributed throughout a wide area network (WAN), namely the internet. It is then kept and analysed by a commercial application on the company's network. Following this, the output is sent back to the person who hired us [7]. This technique of client-server data processing has been in use for most traditional business applications for a considerable period.

However, conventional data centre infrastructures are unable to cope with the rapid rise in the count of web devices and the resulting increase in data density.

IT architects have increasingly switched their emphasis away from the actual data centre and toward the virtual edge of their infrastructures as a function of the transfer of computing systems from the data centre to the area where the data is created. In most cases, data centres should be positioned as close to their customers as feasible; however, where this is not possible, clients should be located as close as possible to the servers. As a result, it was decided that computing resources should be located in the neighborhood of where they are needed, rather than depending on previous generations' understandings of distant computation, such as satellite sites and regional offices, to house computer resources.

In order to collect and analyse data locally, edge computing puts storage and servers on a remote LAN, sometimes requiring as little as a half-rack of equipment in order to carry out the task. For protection against exposure to unfavorable environmental conditions such as temperature fluctuations, moisture incursion, and so on, many computer systems are placed in shielded or hardened enclosures that are designed to withstand impact. It is common for processing to include normalization and analysis of the data stream in order to discover business insight. The results of the analysis are the only ones that are communicated directly to the main data centre.

It can be observed that architecture that is most suitable for one kind of computer activity may not be the framework that is suitable for other types of computing activities. As a result of the emergence of the edge computing architecture, it is now feasible to locate computational resources closer to the data source, ideally at a location that is physically adjacent to the data source. Distant location offices, service centers, data centre hosting, and cloud computing are examples of concepts of distributed computing that have been around for a long time and have a proven track record.

While the move away from the usual centralized computer paradigm may require high levels of management and surveillance, decentralization may prove to be a challenging challenge to accomplish. Large volumes of data are generated and consumed by today's organizations, and edge computing offers a realistic solution to the network issues associated with transmitting that data.

To illustrate, consider the increasing popularity of self-driving vehicles. The only way they'll be able to travel around will be via computerized traffic lights. Real-time data creation and processing will be required for driverless cars. It becomes more obvious how serious the situation is as the number of self-driving automobiles increases. We need a network that is both speedy and responsive in this situation. Edge computing addresses a variety of network issues, including bandwidth, latency, congestion or reliability.

- **Bandwidth:** In computing, bandwidth refers to the amount of data that may be sent through a network at any one time. All networks have a limit on the amount of bandwidth they may use, but wireless networks have significantly tougher limits. This signifies that there is a limit on the amount of data that can be sent over a network or the number of devices that can send data over a network. Increased network capacity can support more devices and data, even if doing so comes with a price that do not provide a solution to other challenges.
- **Latency:** The amount of time it takes for data to be sent from one place to another is referred to as network latency. Despite the notion that data should be able to travel over networks at the velocity of light, physical distances, network congestion, and network failures may cause information to move more slowly than it should. When analytics

and decision-making are postponed, the system's ability to respond in real time is reduced. In the instance of the autonomous automobile, it even resulted in the loss of life.

- **Congestion:** The internet may be solely described as a worldwide "network of networks" Many everyday computer tasks, such as file transfers and rudimentary streaming, may benefit from general-purpose data exchanges via the internet. Because of the sheer number of devices and data being sent over the internet, it's possible that the network may become overburdened, necessitating lengthy information retransmissions and significant levels of congestion. The Internet of Things is rendered worthless if the network fails, since this results in even more congestion and perhaps excludes certain internet users.

When servers and storage are placed near to the source of data creation, edge computing reduces latency and congestion, enabling a much smaller and cheaper LAN to support many more devices. Edge computing is becoming more popular. When data is kept and safeguarded locally, edge servers may perform critical edge analytics or, at the very least, pre-process and limit the data, allowing decision-makers to make decisions in real time prior to uploading findings or merely required data to the cloud or centralized data centre. Figure 3.1 explains the working of cloud computing and edge computing and shows the region where the computation occurs actually.

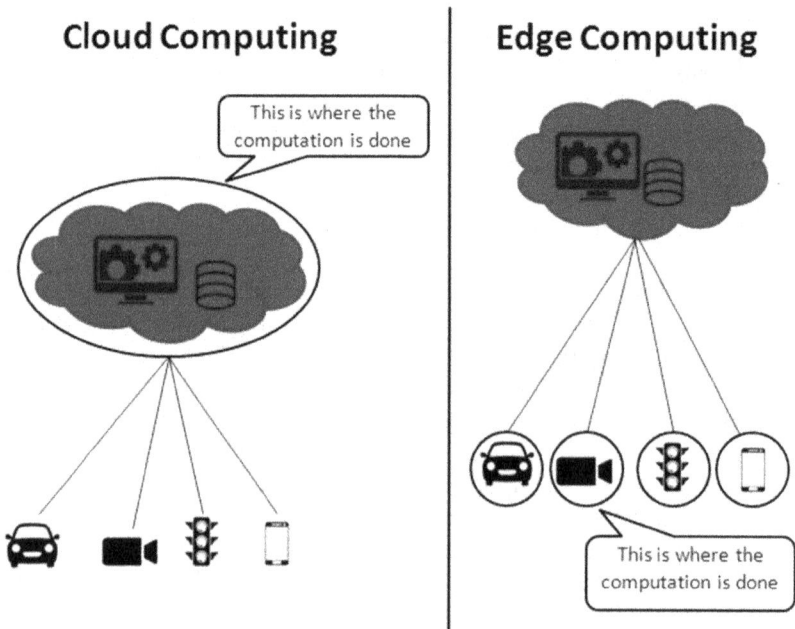

Figure 3.1 Working of cloud computing and edge computing.

3.4 COMPARISON—EDGE COMPUTING, CLOUD COMPUTING AND FOG COMPUTING

"Cloud computing" and "fog computing," two terms that are often used in conjunction with one another when referring to edge computing, are occasionally used synonymously. However, although there are some similarities between the two notions, they are not the same concept and cannot be used frequently in the very same paragraph. Comparing and contrasting the two notions to see how they vary from one another is an interesting exercise:

- A emphasis on the real deployment of power and memory resources in close proximity to the data that is being created are the common traits that run across edge computing, cloud computing, and fog computing, to mention a few of the technologies that are being used today. Understanding the distinctions between these three points of view is a simple issue of observation and comparison. The location of resources has a significant influence on the efficacy of those resources.
- **Edge computing:** In the computing world, "edge computing" refers to the deployment of server and storage facilities close to the site of data generation. In an ideal network, data storage and computing should be positioned at the network's edge. This may be achieved, for example, by placing a small container on top of the turbine that contains many processors and storage for the purpose of collecting and analyzing data collected by sensors. The installation of a small amount of computer and storage capacity to collect and analyze data from a broad variety of railway and rail traffic sensor data is conceivable, for example, at a railway station's ticket counter. Following that, the results of such processing may be transmitted to another data centre for human inspection, archiving, and larger-scale analytics.
  Following that, the results of such processing may be transmitted to another data centre for human inspection, archiving, and larger-scale analytics.
- **Cloud computing:** In addition, there are several places across the globe (regions) where power and memory resources may be deployed, which makes cloud computing incredibly scalable and readily accessible. Well before the era of the Internet of Things, services from cloud service providers were also making the cloud a much more enticing central platform for Internet of Things deployments in the cloud, according to Gartner. Cloud computing provides more than enough resources and services to handle complex analytics, but the nearest geographic cloud facility may be sited hundreds of miles away from the point where data is collected, and connections rely on the same unreliable internet connectivity that is used to support traditional data centers. Although

cloud computing is not intended to be a replacement for traditional data centers, it is often utilized in combination with these facilities. Despite the fact that the cloud has the capacity to move centralized processing closer to the source of data, it is hard for the cloud to reach the network's edges.

- **Fog computing:** Although there are many different options for deploying computer and storage resources, there seems to be no "one-size-fits-all" approach. Where there is no access to a cloud data centre, the edge implementation may be too resource-constrained and/or geographically scattered to be economically sustainable. An emerging idea known as fog computing could be useful as a consequence.

Consequently, fog computing systems may generate massive amounts of sensor or Internet of Things data that cannot be specified by an edge. Advanced technologies, smart cities, and smart utility systems all offer examples of how technology is being used. In a smart city, data may be used to monitor, assess, and enhance public transit, municipal utility systems, and civic services, among other things. Fog computing may operate numerous fog node deployments inside the range of the environment in order to collect, analyze, and interpret information across a big enough region.

Even among technology experts, the terms fog computing and edge computing are often used interchangeably because of their almost identical definitions and designs, and the terms are sometimes used interchangeably when referring to cloud computing.

Starting with the recognition that cloud computing and edge computing are two unique concepts that cannot be swapped for one another, it is vital to understand how they differ [8]. In essence, edge computing is used to handle data that must be processed in a timely fashion, while cloud computing is used to process data which does not need immediate attention. Figure 3.2 explains the architecture in detail.

Edge computing surpasses cloud computing in terms of latency when used in remote places with inadequate or no connectivity to a centralized data centre or server. Edge computing is the ideal solution for the internal storage that is needed at these locations, which may be thought of as a small data centre. Table 3.1 explains the comparison in detail.

Edge computing may also be beneficial to devices that are extremely specialized or intelligent. Despite the fact that these devices look and operate in a similar way to PCs, they are not general-purpose desktops. Because they are intelligent, specialized computers behave in a certain way when they encounter certain other devices. Edge computing, by contrast, has a downside in some industries where rapid responses are required for success.

Edge computing may also be beneficial to devices that are extremely specialized or intelligent. Despite the fact that these devices look and operate in a

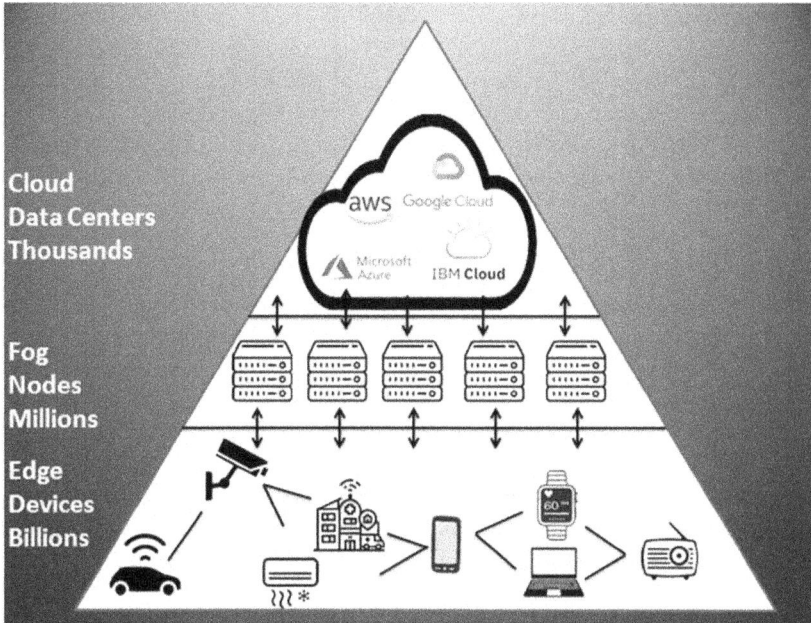

Figure 3.2 Architecture of interaction between Edge, Fog and Cloud.

Table 3.1 Comparison of cloud, fog and edge computing

Parameter	Cloud computing	Fog computing	Edge computing
Data processing site	Cloud server	Inside a fog node or an IoT gateway	Device itself
Objective	Long-term in-depth investigation	Real-time and immediate	Real-time and immediate
Working	Independent	Fog works in integration along with the cloud	It can be defined without cloud
Latency	High	Low	Very low
Security	Best	Better	Good
Architecture	Centralized with regions and availability zones	Hierarchical and flat architecture with several layers that form a network	Separate nodes

similar way to PCs, they are not general-purpose desktops. Because they are intelligent, specialized computers behave in a certain way when they encounter certain other devices. Edge computing, on the other hand, has a downside in some industries where rapid responses are required for success.

3.5 THE BENEFITS OF EDGE COMPUTING

There are a number of significant benefits of this kind of computation that may make the technology attractive in some situations, such as the reduction of bandwidth constraints, the elimination of excessive delay, and the elimination of network congestion:

1. **Autonomy**: Edge computing is advantageous in circumstances where the internet connection is intermittent, or when the bandwidth is limited. The oil rigs, ships on the high seas, and other remote locations are all instances of this. Only when a network connection is available can data be preserved for transmission to a central location. For example, water quality sensors on filtration systems in rural villages may be used to save data for transmission. A significant amount of bandwidth or time is conserved by minimizing the amount of information that has to be sent and by processing data on the local network.

2. **Data sovereignty**: Whenever it comes to transferring massive amounts of data, the problem is more than only technical in nature. When travelling beyond national and regional boundaries, data protection, confidentiality, and other legal concerns may become more difficult to deal with. Edge computing allows data to be retained close to the place of origin while remaining within the constraints of current data sovereignty requirements, such as the General Data Protection Regulation (GDPR), which specifies how data is kept, processed, and exposed. Access to a local processing of raw data helps to protect private information before it has been sent to the cloud or to a central data centre, which may be placed in remote locations across the world.

3. **Edge security**: Finally, edge computing has the ability to make the process of creating and maintaining data security more straightforward. However, despite the fact that cloud providers provide Internet of Things services and expertise in complicated evaluations, organizations are still concerned with the safety and protection of information after it has left their network perimeter. In situations where Internet of Things devices are prone to hacking and other destructive actions, security may be addressed by using computing at the edge in order to allow encryption and protecting the deployment itself against such threats.

3.6 INSTANCES AND ILLUSTRATIONS OF EDGE COMPUTING

In theory, data may be processed and evaluated "in-place" at or close to the network edge adopting edge computing approaches, which allow data to be processed and examined in real time. A powerful way to make use of data that cannot be transferred to a central location at the outset—usually due to the sheer volume of data making such migrations economically difficult, technically tricky, or otherwise infringing specifications, such as data sovereignty—is to store it in a decentralized location. There are several examples of how this phrase has been used in the real world, including:

- **Industrial production:** In order to discover production flaws and improve product quality, one industrial business deployed edge computing to analyze manufacturing operations in real time, with the goal of improving product quality. Environmental sensors were planted throughout the manufacturing plant to evaluate how each manufacturing component is created and maintained, as well as how often the components are kept in stock for each product. These sensors were made possible by edge computing. This allows the corporation to make decisions about its manufacturing plant and operations with increased speed and accuracy as a consequence of the new information technology.
- **Farming:** It could be possible to create an indoor farm that does not need sunlight, soil, or pesticides. Such an approach could make time savings of more than 60%. Sensors enable the firm to keep track of its resource use, nutrition density, and harvesting efficiency, among other things. Crop production algorithms are continually being updated based on the data collected and assessed in order to ensure that crops are harvested in the best possible condition.
- **Optimization of the network:** Edge computing may be used to help enhance network performance by analyzing the performance of users across the internet and then finding the most dependable, low-latency network channel for each user's data to be sent. The use of edge computing for time-sensitive traffic has the effect of acting as a "steering" technique for congestion.
- **Safety at work:** Edge computing, which makes use of on-site cameras, worker safety systems, and other sensors, can assist organizations in monitoring working conditions and ensuring that employees abide to defined safety rules, particularly in remote or dangerous workplaces such as construction sites or oil rigs.
- **Better healthcare:** A significant increase in the amount of patient data collected by gadgets, sensors, and other hospital equipment has occurred in the healthcare industry. Edge computing will have to use automation and machine learning to obtain information, disregard "normal" data, and detect issue data in order for doctors to take quick action to help patients in preventing health crises. This will allow doctors to take immediate actions to assist patients in preventing health crises.
- **Transportation:** On a daily basis, autonomous vehicles consume and create anywhere between 5 TB and 20 TB of data, gathering information about their own location, speed, and other factors, as well as

information about other drivers and vehicles on the road. It is critical that all of this information be gathered and processed while the automobile is going forward. As each autonomous vehicle evolves into a "edge," this necessitates the use of a significant amount of onboard computational capacity. According to the real-world situations, authorities and companies may also benefit from the data in order to better manage existing vehicle fleets and operations.

- **Retail**: Large volumes of data may be obtained via CCTV, stock monitoring, sales data, and other real-time business information, all of which can be used by retail organizations. Edge computing may be used to identify a successful elevated platform or campaign. It can also be used to predict sales and optimize vendor purchasing decisions. Because local retail environments might differ significantly, edge computing can be a cost-effective choice for storing and processing at each location.

3.7 EDGE MAINTENANCE

Edge implementation would be incomplete if it did not also take into consideration edge maintenance:

- Technologies that are primarily concerned with the management of risks as well as the preventive threats should be given serious consideration. Because every device is a network component that may be compromised, there are an overwhelming number of distinct threat vectors for hackers to take advantage of.
- It is necessary to provide access to control and monitoring even if the real data is not available, due to the fact that connectivity is another issue. As a fallback for both connectivity and management reasons, some edge installations make use of a supplementary link as part of their design.
- Edge deployments need remote provision and management due to the fact that they are located in inaccessible and usually hostile areas. The ability to watch what is going on outside of their company and to restrict deployments when necessary is essential for information technology (IT) management professionals.
- It's hard to overstate the significance of physical care and maintenance. The usage of Internet of Things (IoT) devices has been shown to dramatically diminish their usable life when batteries and devices are replaced on a regular basis. It is inevitable that equipment will wear out and that they will need to be fixed or updated at some time. Maintenance must also take into consideration the practical challenges associated with the site.

3.8 OPPORTUNITIES FOR 5G, IoT WITH EDGE COMPUTING

Edge computing is continually evolving and enhancing its functionality via the application of modern techniques. It is expected that edge services would be widely available by 2028, making it the most important development. Edge computing, rather than being situation-specific, is projected to become more prevalent and transform the way the internet is used in the future, offering more abstraction and potential applications for edge technology.

Recent years have seen a rise in the number of edge computing-specific processing, memory, and connectivity devices. In order to increase product interoperability and flexibility, more multi-vendor coalitions are needed. To illustrate this, Amazon Web Services (AWS) and Verizon have collaborated to boost connection at the network's edge.

Increasingly widespread use of new wireless systems, such as 5G and Wi-Fi 6, is expected over the next few years, with a substantial influence on edge installations and usage in the process. As a result of these technologies, previously unimaginable virtualization and automation capabilities, such as greater vehicle autonomy and workload migration to the edge, will be made possible.

Edge computing has become more popular as a result of the growth of the Internet of Things and the huge quantity of data it creates. Although edge computing is still in its early stages, the proliferation of Internet of Things devices will have an impact on its future development. Future ideas include the production of micro modular data centres (MMDCs), which are small data centres that can be moved around. As the name implies, it is a mobile data centre that may be installed in strategic areas, such as around a town or state, to bring computers much closer to the data that they are processing without jeopardising the protection of data being processed.

3.9 CHALLENGES OF EDGE COMPUTING

Although edge computing has the potential to benefit a broad variety of applications, the technology is not without its drawbacks. Furthermore, in addition to conventional network limits, there are a number of significant problems that may have an influence on the deployment of edge computing:

- **Capacity is limited:** The extensibility of cloud computing's resources and services are two of the factors that make edge computing, also known as fog computing, so appealing. An edge architecture may be beneficial, but only if the scope and purpose of the infrastructure are well specified. Even a large-scale solution of edge computing performs a specific role by using limited resources and providing a limited number of services.

- **Connectivity**: Any edge deployment, no matter how tolerant, will need some type of connectivity, even under the most permissive of circumstances. When developing an edge deployment, it is critical to consider what happens if a connectivity is lost at the edge of the network. The independence, artificial intelligence, and elegant failure planning required for successful edge computing are all essential in the case of connection challenges.
- **Security**: When building an edge computing deployment, it's vital to consider elements such as policy-driven configuration enforcement, security in the computing and storage and confidentiality in both the data at rest and in flight, with a particular emphasis on the latter. Even if cloud-based IoT services enable secure communications, if you're constructing an edge location from the ground up, this isn't normally included in the service fee.
- **Data lifecycles**: There is a recurring problem with today's data overload: a significant amount of it is completely meaningless. Think about a medical monitoring equipment, for example: only the problem data is important, and preserving days of normal client records is not essential. Because it depends primarily on short-term data that is not even stored for the long haul, real-time analytics is inefficient. After completing an analysis, it is up to the company to decide which information to preserve and which information to reject. Consequently, companies and regulators must take steps to guarantee that the information they possess is protected.

3.10 CONCLUSION

In the commercial world, edge computing has become increasingly prevalent. However, edge computing is not the only approach available to businesses. When faced with computing challenges, cloud computing is still a decent option for IT vendors and organizations. They may choose to use it in combination with edge computing to provide a more comprehensive solution. In addition, the direction of all data to the edge is not a smart strategy. Accordingly, public cloud providers have started to integrate Internet of Things strategies and tools stacks with edge computing.

As we've discussed in this review, edge computing and cloud computing are not regarded as interchangeable concepts. Indeed, many organizations have embraced edge computing technology, drawn by its promise to overcome some of the basic difficulties connected with cloud computing.

REFERENCES

1. W. Shi, J. Cao, Q. Zhang, Y. Li, and L. Xu, "Edge Computing: Vision and Challenges," *IEEE Internet of Things Journal*, vol. 3, no. 5, pp. 637–646, October 2016.

2. L. Wu, R. Zhang, Q. Li, C. Ma, and X. Shi, "A Mobile Edge Computing-Based Applications Execution Framework for Internet of Vehicles," *Frontiers of Computer Science*, vol. 16, no. 5, January 2022.
3. L. Yuan et al., "CSEdge: Enabling Collaborative Edge Storage for Multi-Access Edge Computing Based on Blockchain," *IEEE Transactions on Parallel and Distributed Systems*, vol. 33, no. 8, pp. 1873–1887, August 2022.
4. N. Abbas, Y. Zhang, A. Taherkordi, and T. Skeie, "Mobile Edge Computing: A Survey," *IEEE Internet of Things Journal*, vol. 5, no. 1, pp. 450–465, February 2018.
5. B. Varghese et al., "Realizing Edge Marketplaces: Challenges and Opportunities," *IEEE Cloud Computing*, vol. 5, no. 6, pp. 9–20, November 2018.
6. S. Wang, Y. Zhao, J. Xu, J. Yuan, and C.-H. Hsu, "Edge Server Placement in Mobile Edge Computing," *Journal of Parallel and Distributed Computing*, vol. 127, pp. 160–168, May 2019.
7. Stephen J. Bigelow, "What Is Edge Computing? Everything You Need to Know," SearchDataCenter, December1, 2021. https://www.techtarget.com/searchdatacenter/definition/edge-computing
8. Shivam Arora, "Edge Computing Vs. Cloud Computing: Key Differences [2022 Edition]," Simplilearn.com, January 16, 2018. https://www.simplilearn.com/edge-computing-vs-cloud-computing-article

Chapter 4

Industrial Internet of Things
IoT and Industry 4.0

Asmita Singh Bisen and Himanshu Payal
Sharda University, Greater Noida, India

CONTENTS

4.1 INTRODUCTION

Today, as a result of rapidly changing technological advances, smart devices can connect to each other. In addition, over the years, the capacity of these devices has decreased, and their depurational capacity and repertory resources have increased. Modern smart devices are bedecked with embedded systems and have the ability to communicate, detect, activate and search, collect, cached and process data in real time. The Internet of Things (IoT) is an innovative and fast-growing technology with a variety of applications, features and services. The IoT allows us to penetrate our required environment and its goals, connect the physical world with the digital world, and "connect people and devices anytime, anywhere, with anything and everyone" [1, 2]. The digitization and automation of industrial production combines ancestral manufacturing and industrial practices with large-scale machine-to-machine communications, IoT, Cyber-Physical Systems (CPS) and many other innovative technologies [3].

4.2 STREAMLINED WORK

Many sophisticated studies have been conducted presenting and analyzing the concepts, characteristics, applications, and key implementation technologies of IoT. They reviewed various IoT phenomena and paradigms and discussed key processing technologies. In addition, the domains of the IoT application are analyzed and grouped in the areas of transport and logistics, health, personal and social, smart environment and futures [4, 5].

In his research [6], he introduced a concentrated "cloud-based" vision for IoT organizations around the world. They likewise examined an IoT outline, as well as key working advances and application areas, and some IoT patterns and scientific categorizations. Moreover, they defeated clear difficulties and future patterns in the IoT business and introduced an information examination and contextual analysis on the Aneka/Azure cloud stage. Finally, the requirement for combination of the remote sensor organization, the Internet and disseminated registering was featured [7, 8].

In an effort to clarify the need for high-end IoT hardware, [9] conduct research on existing IoT media opportunities, challenges and supporting technologies. Advanced centralized solutions for implementing IoT applications are provided, as well as a complete analysis of problems and support technologies in IoT-centric software development. IoT is a dynamic and global ubiquitous computing methodology which enables the quality of life and ease of end user.

4.3 INTERNET OF THINGS

The Internet of Things (IoT) is one of the most widely used terms in the development of modern computing. Although often referred to as technology, it is more accurately described as a platform for integrating objects (through recognition) so that the information collected about them can be used to analyze, interpret, define and execute this and other related information [10, 11].

The term Industry 4.0 comes from an undertaking in the German government's cutting-edge strategy. Such an undertaking upholds robotization. This depends on innovative ideas for the Internet of Things (IoT), which empowers the production line representing things to come [12, 13].

Since then, companies have presented solutions to this concept, with the help of governments, especially Europe (especially Germany), but also countries like the United States, Japan and China, indicating that this was an industrial and strategic era.

In this, field gear, machines, other items comprise a digital actual framework that independently trades data, starts activities and drives each other freely. Changing production lines into a brilliant climate that overcomes any issues between this present reality and the computerized world [14]. The solid tendency of the electrical and various leveled universes of plant mechanization will move toward savvy industrial facility networks that empower dynamic re-designing cycles and the capacity to respond deftly to interruptions and disappointments.

The vital parts of this is that it can be characterized by three ideal models: a brilliant item, a savvy machine and high-level administrator. The directing thought behind the Smart Product is to make the section a functioning piece of the framework. Items get a memory where organization information and prerequisites are put away straightforwardly as isolated building plans. The smart machine worldview portrays the cycle by which machines become Cyber-Physical Manufacturing Systems (CPMS). Expanded Operator, the third worldview referenced above, centers around the mechanical help of laborers in a perplexing climate of profoundly measured creation frameworks.

4.4 THE INDUSTRIAL INTERNET OF THINGS

Modern IIoT is a particular kind of IoT that spotlights applications in the cutting-edge industry and savvy production. With regards to Industry 4.0, IIoT should be visible as a complicated framework with numerous frameworks and gadgets. Specifically, IIoT consolidates a few significant cutting-edge innovations to make a framework that works more productively than a simple combination of its parts [15, 16].

With authentic assets, network innovation, applications, sensors, programming, media and storehouses, IIoT conveys arrangements and capacities that empower and control the business cycles and resources. IIoT administrations and applications give basic answers for better preparation, booking and the board of items and frameworks [17, 18].

Besides, through the different associated gadgets that can and do speak with one another and with additional brought together regulators, IIoT will decentralize investigation and navigation, generating ongoing input and reactions [19]. Therefore, both enterprises' general accessibility and supportability and their presentation have improved, creation has been accelerated, item time-to-showcase is more limited at low rates, and their general exhibition is enhanced, arriving at huge capacities at a level up until recently never accomplished financial development and efficiency effectiveness [20, 21].

4.5 INDUSTRY 4.0

In recent years, the global business environment has changed dramatically due to the advent of technology, innovation and innovation. The fourth variant (Industry 4.0) is its newest application and function, multiple integrated devices, as well as its new product [22].

It is controlled by the accompanying four innovation groups:

1. PC information, power and availability,
2. examination and knowledge,
3. human–machine cooperation, and
4. computerized-to-actual transformation [23].

Likewise it joins the force of conventional enterprises with trend-setting innovations that empower savvy items to be incorporated into interweaved advanced and actual cycles. These cycles are associated with one another [24, 25].

The primary plan standards for different parts of Business 4.0 are:

1. reconciliation,
2. virtualization,
3. decentralization,
4. in the nick of time stacking,
5. administration accessibility [26].

Industry 4.0 addresses the change from "concentrated" to "decentralized". Because of the circulation of knowledge, clever gadgets are created in free cycle control frameworks. A significant new part of endlessly fabricating processes is the mix of the genuine and virtual universes, where fabricating hardware both produces items and communicates with them. Accordingly, ventures, creation expenses, and plans of action become more astute, prompting the development of "savvy" industrial facilities [27].

4.6 KEY TECHNOLOGIES FOR THE INDUSTRIAL INTERNET OF THINGS

IIoT support is intangible and combines a wealth of epic editing, including IoT computing, comprehensive evaluation information, artifact information, real-world structural mechanisms, and extended operations.

4.6.1 Blockchain technology

Considering the related handling plant circumstance, Internet of Things contraptions help steady data collection and incitation. As the most fundamental part of IIoT, these contraptions track the creation line assets across the globe [28]. The devices in a totally related IIoT framework are sent across all the assembling plant workplaces, going from stockrooms to creation workplaces and dispersion focuses. Regardless, the configuration, sending, checking, and upkeep of these contraptions is a troublesome errand and require extraordinarily qualified specific staff.

4.6.2 Cloud computing

The dramatic development of information in IIoT requires a system that requires a first-order system to manage, manage, display, and store information. A distributed distribution system provides statistics, membership, and control over all online jobs and production systems. All content and accessories attached to it are attached to the remains. Cloud sites are arranged as private, public (only security for cloud-based clients) or ivory (using both types of auxiliaries) [29]. Since the underpinning of waiter ranches and the enlistment of specific staff require high spending thusly confidential cloud organization models are not a reasonable decision for as of late members as well as nearly nothing and medium-level undertakings. However, large and well-established overall undertakings favor the sending of private fogs to ensure the wellbeing, security, and assurance and adjust to current covert work for high ground.

4.6.3 Artificial intelligence and cyber physical systems

Man-made information advances guarantee that the IIoT construction ought to run uninhibitedly and magnificently to limit the thoughtful intercessions and further encourage efficiency. The AI movements make IIoT free by utilizing complex AI advancements, for instance, multi-master frameworks and conversational AI [30]. In addition, the data is implanted at layers in the Industrial Internet of Things structures from sensors to gadgets to edge servers and cloud server farms by empowering different chase, optimization, and guess calculations. To confine human undertakings and mediations, IIoT structures attract different advanced certifiable frameworks like gathering designs and present-day robots. The substance of computerized genuine structures lies in locally accessible embedded IoT contraptions which empower different sensors and actuators to work in the state of the art conditions.

4.7 INTELLIGENT MANUFACTURING IN THE CONTEXT OF INDUSTRY 4.0

Fabricating involves a central modern part which fundamentally affects individuals' vocation and a country's economy. Fully intent on upgrading the general creation, efficiency and item-quality administration all through the different phases of the lifecycle of items, IoT offers applications and administrations which incorporate high-level observing and following, execution and practicality enhancement and human–machine connection [31]. Subsequently, it makes sense that IoT can give a ton of answers for the assembling space which is described by its intricacy and expansiveness of utilizations, its different Cyber Physical Systems and its assembling activity and the board philosophies.

Clever assembling utilizes the consolidated knowledge of individuals, cycles and machines in order to increment creation. It offers remarkable answers for the location and checking of possible harm, glitches and breakdowns. Additionally, it upgrades control and the executives, further develops practicality and accessibility and streamlines asset the board and sharing. Moreover, it applies state-of-the-art advancements to different customary frameworks, administrations and items [32]. Thus, clearly keen assembling definitely affects the general capacity and financial aspects of undertakings and will make ready for the headway of current ventures [33].

Shrewd assembling targets growing continuous, independent and human-like insightful dynamic frameworks that decrease the requirement for human contribution and intercession. In order to achieve this, man-made brainpower, AI, hereditary calculations and other cutting-edge innovations, approaches and methods are utilized. This reality contains a significant distinctive element between canny assembling and customary assembling [34]. In any case, the objective of both assembling spaces continues as before, or, at least, to fulfill clients' prerequisites and market needs as well as amplifying benefits while at the same time limiting conceivable expense and waste.

4.8 OPEN RESEARCH ISSUES

Clearly with regards to the use of these can upgrade and change the ongoing enterprises and yield a great deal of advantages because of its cutting-edge innovations, applications and administrations. It is likewise fundamental to highlight that IoT does not focus solely on changing businesses and expanding their efficiency. In addition, it enhances the central motivation behind ventures and relieving the shortcomings brought about by inheritance frameworks. Thus, it ought to be viable with existing gadgets, frameworks and foundations and have the option to implant knowledge into them [35]. Accordingly, undertakings that are going through advanced change will be worked with to take on and execute IoT and take advantage of its various advantages and arrangements without having to straightforwardly put resources into absolutely spic and span hardware as cost would far offset the prompt advantages. In any case, for this to be completely carried out and for IoT to be embraced and completely used by businesses and undertakings, a great deal of difficulties and open issues ought to be investigated.

Many elaborate investigations, which break down indispensable IoT difficulties, coordination and execution issues and open exploration issues, have been directed. All the more explicitly, open exploration issues, including normalization exercises tending to and organizing as well as security and protection were examined by [36]. Key IoT challenges, for example, interoperability and normalization, information and data secrecy, encryption and protection, the naming and character of the board, IoT greening as well as article and organization security were depicted [37].

Correspondence and recognizable proof advancements, dispersed framework innovations and insight and accentuated security issues, for example, information secrecy, protection and trust were aspects of the primary examination challenge that were investigated [38]. Information the executives and mining, security and protection were viewed as the fundamental difficulties which undertakings face in IoT advancement zeroed in on security and protection moves in IIoT and their weakness to an assortment of cyberattacks [39, 40].

Many elaborate investigations, with respect to imperative difficulties, combination and execution issues and the open exploration issues of Industry 4.0, have been done. These have identified a few difficulties and principal issues in different segments that happen all through the execution of the advancements of the IoT [41, 42].

The recognized areas in their examinations were:

1. Shrewd direction and exchange component,
2. Fast modern remote organization conventions, fabricating explicit large information and its scientific,
3. Framework demonstrating and investigation,
4. Digital security and
5. Modularized and adaptable actual ancient rarities,
6. Investment issues.

A few producers and endeavors wonder whether or not to proceed like this because of specific worries and obstructions. These incorporate vulnerabilities about monetary advantages, the absence of systems of planning across various authoritative units, missing ability, abilities and capacities, wavering to go through extremist change and concerns with respect to the outsider suppliers' security [43, 44].

To summarize, in the light of the previously mentioned examinations, the most huge and normal difficulties and open exploration issues which businesses and undertakings ought to know about are:

a. Availability, unwavering quality, portability and other QoS models;
b. Security, protection and privacy of information;
c. Interoperability and versatility;
d. Fault resilience and usefulness for wellbeing;

   e. Management of tasks, assets, energy and information;

   f. Networking tending to and ID;

   g. Architecture, conventions and normalization exercises

   h. Trouble in planning activities across various hierarchical units, like exploration and improvement such as Research and Development (R&D), IT, assembling, deals, and money divisions, because of unfortunate cooperation between them; [45]

   i. Uncertainty about adopting as opposed to rethinking and the absence of information about specialist co-ops;

   j. Concerns about online protection while including outsider innovation/programming and execution suppliers;

   k. Concerns about information proprietorship while working with outsider suppliers;

   l. Challenges with coordinating information from unique sources to empower Industry 4.0 applications.

4.9 APPLICATION DOMAINS

Albeit the new advances in universal registering and the possibilities presented by the IoT render the improvement of countless applications practical, advances are already visible in a couple of areas [46]. IoT applications target further developing quality for the end-client local area and supporting framework and universally useful tasks [47].

The intricacy and the size of the issue to be addressed, as well as the particularities, prerequisites and attributes of the particular areas in which they will be carried out, ought to be thought about as there is no "one-size-fits-all" arrangement. Also, they ought to be planned circumspectly in order to fulfill different goals and prerequisites while at the same time expanding the nature of involvement and nature of administration levels [48–50]:

   i. Ecological space: This includes applications that secure, screen and foster every single regular asset, natural administration administrations, energy the board, reusing, farming and so on

   ii. Industrial area: Applications of this area include monetary or business exchanges between ventures, associations and different elements. Also, they allude to assembling, planned operations, banking, monetary legislative specialists and so forth

   iii. Social area: This includes applications with respect to the turn of events and the incorporation of social orders, urban communities, and individuals as well as administrative administrations toward residents and other society structures [51].

There are a variety of applications, including the following:

4.9.1 Healthcare

The clinical area is quite possibly the earliest business to have adopted the IoT and in which it can be seen to have had the most significant impact. Advances in this area can open new doors, administrations and applications to work on the medical care and clean area. In addition, IoT stages and administrations improve current living arrangements and work with the acknowledgment of ubiquitous medical services vision that is "medical services to anybody, whenever, and anyplace by eliminating area, time and different restrictions while expanding both the inclusion and nature of medical care" [52]. This reality will improve and computerize the method involved with gathering information, thereby producing a phenomenal amount of information that can be used to pursue additional logical and clinical investigations. With the end goal of forestalling the beginning of medical issues all the more proficiently, IoT advancements work on persistent therapy and prosperity.

4.9.2 Smart cities

Rapid metropolitan development is now overwhelming the current foundation and utilities and featuring the requirement for more manageable metropolitan preparation and public administrations. The applications and administrations are being taken advantage of for these new necessities to be fulfilled and the cultural changes proportionate with this rapid development to be answered. In addition, autonomous city and home organizations will be keen and fit for detecting, observing and adjusting to ecological pliancy as well as responding to human exercises. All the more explicitly, shrewd innovations and gadgets are interconnected and, subsequently, they can improve and upgrade the quality and way of life for city tenants as well as guarantee that their fundamental administrations are given [53]. Moreover, by making shrewd urban communities that mix the generally settled city administrations and utilities that occupants cooperate with consistently, advancing the use of city foundation, assets and offices and upgrading city inhabitants' life quality.

4.9.3 Smart environments

By using completely interconnected mechanical gadgets and implanted frameworks, IoT targets invade our regular climate and its items and make better approaches to associate with these brilliant conditions [54].

Through the involvement of IoT in a blend with mechanized programming specialists for ongoing following and observing, savvy conditions become a mechanical environment of different interconnected gadgets. These brilliant gadgets can safely convey and cooperate as well as recover, cycle, store and trade information on a continuous basis. By incorporating these heterogeneous information into applications, the variation cycle to inhabitants' and natural persistently changing requirements is worked with. Thus, overall, their prerequisites are speedily and sufficiently met. In addition, IoT applications in this area target working on the ongoing natural wellbeing by lessening and relieving the likely effect of harm and catastrophe. IoT advancements permit the improvement of creative constant checking and dynamic emotionally supportive networks and applications with regard to ecological issues, for example, early expectation and recognition of cataclysmic events, weather patterns and so on.

4.9.4 Industry

A particular classification of IoT centers around its cases in current ventures and savvy production. Being an intricate arrangement of a wide assortment of systems is thought of. Besides, it includes a critical part to modern space and is firmly connected with the fourth modern transformation (Industry 4.0). It joins a few creative key innovations in order to deliver a framework that capacities more successfully than the amount of its parts.

4.10 CHALLENGES

The variegated and labyrinthine nature of its structure has faced numerous challenges, in areas such as safety and security, adaptability, heterogeneity and the quality of real tools.

To cope with these difficulties, efficient information the executives models are required. These information the board models ought to be fit for efficiently taking care of the tremendous measure of crude information generated by gadgets [55, 56]. These models should in like manner give the data to the board administrations with fast data taking care of, trustworthy and secure data accumulating, recuperation and speedy data flow.

4.11 CONCLUSION

IoT is an inventive and rapidly developing innovation which offers different novel applications, administrations and arrangements and connects the physical to the advanced world. In addition, it targets changing the ongoing businesses into shrewd ones using the powerful organization of interconnected gadgets. Improving their activity and usefulness, expanding their efficiency and lessening their expenses and waste are among the many advantages and benefits that undertakings can acquire by utilizing IoT. Further, ventures that completely embrace IoT will be in front of their rivals, become more agile, adjust to the ceaselessly evolving market, make results greater that fulfill clients' necessities and prerequisites.

Additionally, with regards to Industry 4.0, IoT, and all the more explicitly IIoT, can be used in blend with other creative advances, for example, huge information, distributed computing, CPSs and so on to improve and change the ongoing assembling frameworks into smart ones. It considers machines to become autonomous substances that can gather and break down information and offer guidance upon it without requiring any human intercession as it presents self-viability, self-advancement, and self-perception. It tries to adapt effectively to the worldwide serious nature of the present business sectors and enterprises in accordance with the clients' steadily changing necessities and prerequisites.

In spite of the fact that IoT offers an abundance of answers for ventures as well as a large number of contemporary and high-level applications and administrations, it is currently at an early phase of improvement, reception and execution. Accordingly, for the different momentum moves and open issues to be experienced and settled, further exploration ought to be done. All things considered, the total execution and brief reception of IoT alongside fitting usage of its original advances, applications and administrations cannot further develop life quality; however, it can likewise yield huge individual, proficient and financial open doors and advantages sooner rather than later.

REFERENCES

1. Perera, C., Liu, C. H., & Jayawardena, S. (2015). The emerging internet of things marketplace from an industrial perspective: A survey. *IEEE Transactions on Emerging Topics in Computing*, 3(4), pp. 585–598.
2. Vermesan, O., Friess, P., Guillemin, P., Gusmeroli, S., Sundmaeker, H., Bassi, A., … & Doody, P.(2011). Internet of things strategic research roadmap. *Internet of Things-Global Technological and Societal Trends*, 1(2011), pp. 9–52.
3. Li, J., Huang, Z., & Wang, X. (2011, May). Notice of Retraction Countermeasure research about developing Internet of Things economy: A case of Hangzhou city. In *2011 International Conference on E-Business and E-Government (ICEE)*, pp. 1–5. IEEE.
4. Atzori, L., Iera, A., & Morabito, G. (2010). The internet of things: A survey. *Computer Networks*, 54(15), pp. 2787–2805.
5. Sundmaeker, H., Guillemin, P., Friess, P., & Woelfflé, S. (2010). Vision and challenges for realising the internet of things. *Cluster of European Research Projects on the Internet of Things, European Commision*, 3(3), pp. 34–36.
6. Gubbi, J., Buyya, R., Marusic, S., & Palaniswami, M. (2013). Internet of things (IoT): A vision, architectural elements, and future directions. *Future Generation Computer Systems*, 29(7), pp. 1645–1660.
7. Al-Fuqaha, A., Guizani, M., Mohammadi, M., Aledhari, M., & Ayyash, M. (2015). Internet of things: A survey on enabling technologies, protocols, and applications. *IEEE Communications Surveys & Tutorials*, 17(4), pp. 2347–2376.
8. Lampropoulos, G., Siakas, K., & Anastasiadis, T. (2019). Internet of things in the context of industry 4.0: An overview. *International Journal of Entrepreneurial Knowledge*, 7(1), pp. 4–19. doi: 10.2478/ijek-2019-000.

9. Ngu, A. H., Gutierrez, M., Metsis, V., Nepal, S., & Sheng, Q. Z. (2016). IoT middleware: A survey on issues and enabling technologies. *IEEE Internet of Things Journal*, 4(1), pp. 1–20.
10. Atzori, L., Iera, A., & Morabito, G. (2010). The internet of things: A survey. *Computer Networks*, 54(15), pp. 2787–2805.
11. Vermesan, O., Friess, P., Guillemin, P., Gusmeroli, S., Sundmaeker, H., Bassi, A., ... & Doody, P.(2011). Internet of things strategic research roadmap. *Internet of Things-Global Technological and Societal Trends*, 1(2011), pp. 9–52.
12. Lee, J., Bagheri, B., & Kao, H. A. (2015). A cyber-physical systems architecture for industry 4.0-based manufacturing systems. *Manufacturing Letters*, 3, pp. 18–23.
13. Haddara, M., & Elragal, A. (2015). The Readiness of ERP systems for the factory of the future. *Procedia Computer Science*, 64, pp. 721–728.
14. Weyer, S., Schmitt, M., Ohmer, M., & Gorecky, D. (2015). Towards Industry 4.0-Standardization as the crucial challenge for highly modular, multi-vendor production systems. *IFAC-Papers on Line*, 48(3), pp. 579–584.
15. Lu, Y. (2017). Industry 4.0: A survey on technologies, applications and open research issues. *Journal of Industrial Information Integration*, 6, pp. 1–10.
16. Lampropoulos, G., Siakas, K., & Anastasiadis, T. (2018). Internet of things (IoT) in industry: Contemporary application domains, innovative technologies and intelligent manufacturing. *International Journal of Advances in Scientific Research and Engineering*, 4(10), pp. 109–118.
17. Gilchrist, A. (2016). *Industry 4.0: The industrial internet of things.* New York, NY: Apress. ISBN: 1484220463.
18. Bi, Z., Da Xu, L., & Wang, C. (2014). Internet of things for enterprise systems of modern manufacturing. *IEEE Transactions on Industrial Informatics*, 10(2), pp. 1537–1546.
19. Şen, K. Ö., Durakbasa, M. N., Baysal, M. V., Şen, G., & Baş, G. (2018). Smart factories: A review of situation, and recommendations to accelerate the evolution process. In *The International Symposium for Production Research*, pp. 464–479. Springer, Cham.
20. Schmidt, R., Möhring, M., Härting, R. C., Reichstein, C., Neumaier, P., & Jozinović, P. (2015, June). Industry 4.0-potentials for creating smart products: empirical research results. In *International Conference on Business Information Systems*, pp. 16–27. Springer, Cham.
21. Zhou, K., Liu, T., & Zhou, L. (2015). Industry 4.0: Towards future industrial opportunities and challenges. In *2015 12th International Conference on Fuzzy Systems and Knowledge Discovery (FSKD)*, pp. 2147–2152. IEEE.
22. Lampropoulos, G., Siakas, K., & Anastasiadis, T. (2018). Internet of things (IoT) in industry: Contemporary application domains, innovative technologies and intelligent manufacturing. *International Journal of Advances in Scientific Research and Engineering*, 4(10), pp. 109–118.
23. Wee, D., Kelly, R., Cattel, J., & Breunig, M. (2015). Industry 4.0-how to navigate digitization of the manufacturing sector. McKinsey & Company, 58.
24. Schmidt, R., Möhring, M., Härting, R. C., Reichstein, C., Neumaier, P., & Jozinović, P. (2015, June). Industry 4.0-potentials for creating smart products: empirical research results. In *International Conference on Business Information Systems*, pp. 16–27. Springer, Cham.

25. Geissbauer, R., Vedso, J., & Schrauf, S. (2016). Industry 4.0: Building the digital enterprise. Retrieved from: https://www.pwc.com/gx/en/industries/industries-4.0/landing-page/industry-4.0-building-your-digital-enterprise-april-2016.pdf

26. Hermann, M., Pentek, T., & Otto, B. (2016, January). Design principles for industrie 4.0 scenarios. In *2016 49th Hawaii international conference on system sciences (HICSS)*, pp. 3928–3937. IEEE.

27. MacDougall, W. (2014). *Industrie 4.0: Smart manufacturing for the future.* Germany Trade & Invest.

28. D. Miller, Blockchain and the internet of things in the industrial sector, *IT Professional* (2018) 15–18 (2018).

29. Khan, W. Z., Rehman, M. H., Zangoti, H. M., Afzal, M. K., Armi, N., Salah, K. (2020). Industrial Internet of things: recent advances, enabling technologies and open challenges. *Computers & Electrical Engineering* 81, p. 106522.

30. Zhong, R. Y., Xu, X., Klotz, E., & Newman, S. T. (2017). Intelligent manufacturing in the context of industry 4.0: A review. *Engineering*, 3(5), pp. 616–630.

31. Davis, J., Edgar, T., Porter, J., Bernaden, J., & Sarli, M. (2012). Smart manufacturing, manufacturing intelligence and demand-dynamic performance. *Computers & Chemical Engineering*, 47, pp. 145–156.

32. Francis, D., & Bessant, J. (2005). Targeting innovation and implications for capability development. *Technovation*, 25(3), pp. 171–183.

33. Parwez, M. S., Rawat, D. B., & Garuba, M. (2017). Big data analytics for user-activity analysis and user-anomaly detection in mobile wireless network. *IEEE Transactions on Industrial Informatics*, 13(4), pp. 2058–2065.

34. Kumar, S. (2002). *Intelligent Manufacturing Systems.* B.I.T. Mesra, Ranchi, India. Retrieved from https://pdfs.semanticscholar.org/fc60/02a47a1a9b312bda871b47569edeeb8545f9.pdf

35. Küsters, D., Praß, N., & Gloy, Y. S. (2017). Textile learning factory 4.0–preparing Germany's textile industry for the digital future. *Procedia Manufacturing*, 9, pp. 214–221.

36. Atzori, L., Iera, A., & Morabito, G. (2010). The internet of things: A survey. *Computer Networks*, 54(15), pp. 2787–2805.

37. Khan, R., Khan, S. U., Zaheer, R., & Khan, S. (2012, December). Future internet: the internet of things architecture, possible applications and key challenges. In *2012 10th international conference on frontiers of information technology*, pp. 257–260. IEEE.

38. Miorandi, D., Sicari, S., De Pellegrini, F., & Chlamtac, I. (2012). Internet of things: Vision, applications and research challenges. *Ad Hoc Networks*, 10(7), pp. 1497–1516.

39. Lee, I., & Lee, K. (2015). The Internet of Things (IoT): Applications, investments, and challenges for enterprises. *Business Horizons*, 58(4), pp. 431–440.

40. Sadeghi, A. R., Wachsmann, C., & Waidner, M. (2015, June). Security and privacy challenges in industrial internet of things. In *2015 52nd ACM/EDAC/IEEE Design Automation Conference (DAC)*, pp. 1–6. IEEE.

41. Wang, S., Wan, J., Li, D., & Zhang, C. (2016). Implementing smart factory of industrie 4.0: an outlook. *International Journal of Distributed Sensor Networks*, 12(1), 3159805.

42. Vaidya, S., Ambad, P., & Bhosle, S. (2018). Industry 4.0 – a glimpse. *Procedia Manufacturing*, 20, pp. 233–238.

43. Lu, Y. (2017). Industry 4.0: A survey on technologies, applications and open research issues. *Journal of Industrial Information Integration*, 6, pp. 1–10.
44. Bauer, H., Baur, C., Mohr, D., Tschiesner, A., Weskamp, T., Alicke, K., & Wee, D. (2016). *Industry 4.0 after the initial hype–Where manufacturers are finding value and how they can best capture it.* McKinsey Digital.
45. Küsters, D., Praß, N., & Gloy, Y. S. (2017). Textile learning factory 4.0 – preparing Germany's Textile industry for the digital future. *Procedia Manufacturing*, 9, pp. 214–221.
46. Atzori, L., Iera, A., & Morabito, G. (2010). The internet of things: A survey. *Computer Networks*, 54(15), pp. 2787–2805.
47. Akpakwu, G. A., Silva, B. J., Hancke, G. P., & Abu-Mahfouz, A. M. (2017). A survey on 5G networks for the internet of things: Communication technologies and challenges. *IEEE Access*, 6, pp. 3619–3647.
48. Gubbi, J., Buyya, R., Marusic, S., & Palaniswami, M. (2013). Internet of things (IoT): A vision, architectural elements, and future directions. *Future Generation Computer Systems*, 29(7), pp. 1645–1660.
49. Hajrizi, E. (2016). Smart solution for smart factory. *IFAC-PapersOnLine*, 49(29), pp. 1–5.
50. Sundmaeker, H., Guillemin, P., Friess, P., & Woelfflé, S. (2010). Vision and challenges for realising the internet of things. *Cluster of European Research Projects on the Internet of Things, European Commision*, 3(3), pp. 34–36.
51. Miorandi, D., Sicari, S., De Pellegrini, F., & Chlamtac, I. (2012). Internet of things: Vision, applications and research challenges. *Ad Hoc Networks*, 10(7), pp. 1497–1516.
52. Lee, C. P., & Shim, J. P. (2009). Ubiquitous healthcare: Radio frequency identification (RFID) in hospitals. In *Handbook of research on distributed medical informatics and e-health*, pp. 273–281. IGI Global.
53. Zanella, A., Bui, N., Castellani, A., Vangelista, L., & Zorzi, M. (2014). Internet of things for smart cities. *IEEE Internet of Things Journal*, 1(1), pp. 22–32.
54. Weiser, M., Gold, R., & Brown, J. S. (1999). The origins of ubiquitous computing research at PARC in the late 1980s. *IBM Systems Journal*, 38(4), pp. 693–696.
55. Javed, F., Afzal, M. K., Sharif, M., & Kim, B.-S. (2018). Internet of things (iot) operating systems support, networking technologies, applications, and challenges: A comparative review, *IEEE Communications Surveys & Tutorials*, 20(3), pp. 2062–2100.
56. Zikria, Y. B., Kim, S. W., Hahm, O., Afzal, M. K., & Aalsalem, M. Y. (2019). Internet of things (iot) operating systems management: Opportunities, challenges, and solution, *Sensors*, pp. 1–10.

Chapter 5

Denial of Service Attacks in the Internet of Things

Aditi Paul

Banasthali Vidyapith, Rajasthan, India

S. Sinha

CHRIST(Deemed to be University), Bangaluru, India

CONTENTS

5.1 INTRODUCTION

The maiden years of the Internet of Things (IoT) mainly included communications between machines, also known as machine-to-machine (M2M) associations [1]. However, over time this concept has developed to incorporate the involvement of humans, so much so that our dependence on its services [2] continues to grow with time. Today our earth is packed with gazillions of computing devices and sensors that perpetually sense, pick up, gather, aggregate, and scrutinize the personal information of most of its population. This information encompasses our address and whereabouts, contacts list, browsing habits patterns, and sensitive and detailed attributes about our health [3]. Hearing, collecting, and spreading such intimate personal data on computer devices is primarily promoted in the aspiration of

DOI: 10.1201/9781003407300-5

comfort; the idea is that more ingenious devices can better comply with our needs and desires and even circumstances (e.g., a thermostat senses its environment and alters the temperatures taking a cue from our location, what time of the day or what day of the week or what time of year it is, etc.). These also tackle crises such as fire or burglary. Unfortunately, this level of installation increases security costs. Also, it imposes specific challenges [4] concerning privacy: personal, confidential information, if available to ill-intended agents, can havoc damage our property, dignity, and our safety.

The relevance of computing devices having such disastrous inadequacies is no longer limited to research papers and academic studies. Still, it is, in fact, a sad reality with a plethora of news snippets about multiple devices succumbing to daily security breaches. And the fact that these attacks are not so difficult to coordinate does not help the case. There have been many demonstrations of such invasions. The attacker can inject [5] malicious code directly into a wearable device through an editing interface and then capture sensitive user data. Attacks on critical medical equipment, for instance, Implantable Cardiac Defibrillators (ICD) and wireless monitors, can be life-threatening for the patient. The industrial and urban framework has also seen an increase in recent attacks. As many of today's premium automobiles use electronic and embedded devices, the attacker may gain control the vehicle [6] by influencing the Electronic Control Unit (ECU). This would be a significant driver safety hazard.

The Distributed Denial of Service Attack (or DDoS attack) [7] poses significant risks to the performance of IoT applications and is amongst the leading virtual threats on the internet. They occur because of the simplicity of their execution, which is further aided by the fact that certain companies offer such attacks as a service [8]. These attacks are much feared because they can overpower their target and defeat the accessibility of their victim within a matter of seconds. Recent reports suggest [9] a growing trend in the scope of DDoS attacks. This has led to incurring financial losses and also impacted millions of users worldwide. GitHub's server attack in 2018 is one of the most devastating DDoS attacks ever recorded. This attack was facilitated by targeting a vulnerable region of the application layer protocol. It was designed to supply 129 million applications per second and reach a total capacity of 1.35 Tbps [10], closely following the most notable attack in 2016 [11], which involved 1.2 Tbps.

This chapter considers various challenges, methods, and practices for IoT security. Based on these, two taxonomies are designed for security attacks targeted at IoT systems. The first taxonomy introduces attacks on four-tiered infrastructures of IoT (Perception Layer, Network Layer, Middleware Layer, Application Layer). Based on this taxonomy, a systematic analysis of security and privacy threats throughout the different layers of IoT is done. The chapter also examines the types of DDoS attacks, how they work, their impact on the IoT ecosystem, and other available protection mechanisms.

5.2 IoT ARCHITECTURE

The Internet of Things (IoT) does not have a fixed architecture. However, the abstraction of layers can be viewed as consisting of either three or four layers. The three-layer architecture [12] comprises an application layer, a network layer, and a perception layer. Since IoT regards both objects and devices as a thing, these are complex. The heterogeneous devices on the internet have their constraints, and complexities grow when these devices are linked to providing services to the objects. To address this issue, a middle layer [13] is introduced to create an abstraction between technology and applications/ services. The middleware layer connects the network layer to the application layer through cloud storage. This means the features of the network layer are stored in the cloud and accessed by the application layer. This reduces the complexity and frees up the Application layer's load. This layer also provides APIs [14] to the Application layer. Thus, the middleware layer supports data-processing capability by simplifying computation. Each layer (Figure 5.1) is discussed briefly, along with the technology used in each layer:

> **Application layer**: The application layer is the service layer that con-
> nects to the objects and provides services. These services include smart
> homes, innovative health, automated vehicles, etc. This layer works

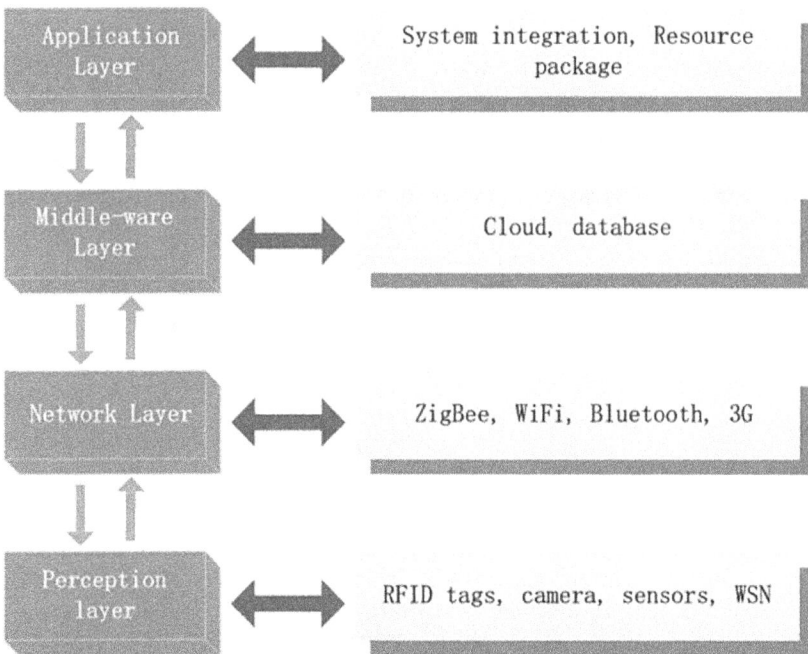

Figure 5.1 Four-layer IoT Architecture.

on formatting and the presentation of data to the user. HTTP is the primary service protocol in this case, but it requires substantial overheads. The resource-constrained devices cannot cope with these high overheads, and hence various lightweight protocols are proposed in Application layers. Constrained Application Protocol (CoAP), Message Queue Telemetry Transport (MQTT), etc., one crucial aspect of the application layer is quality of service (QoS). However, the handling of sensitive information is challenging, and exploitation through the actions of intruders, such as unauthorized access to data, data modifications, etc. The vulnerabilities in application layers create paths for attackers to exploit sensitive data.

Middleware layer: The role of the middleware [15] layer is to provide service management, data storage, and service composition. This layer receives data from the network layer, stores and processes data on the cloud, and provides an application interface to the upper layer. The middleware layer contributes to a more robust data processing and repository. However, the vulnerabilities of cloud storage are one of the challenges in this layer. This, in turn, affects the quality of service in the application layer.

Network layer: This layer is responsible for communication between IoT devices and also for data exchange. Sensor nodes send data from the lower layer, routed through this layer. The Routing Protocol for Low Power and Lossy Networks (RPL) is used, which is suitable for constrained devices. The other protocols involved are ZigBee, Wi-Fi, Bluetooth, and 3G. The network layer is prone to attacks like Denial of Service and eavesdropping [16], which violates the authenticity and availability of data/communication.

Perception layer: The perception layer [17] is the physical layer that identifies objects and collects information. This layer is responsible for converting digital data into signals. This edge layer consists of a sensor-forming Wireless Sensor Network (WSN). These sensors are constrained in energy and power consumption. The deployment of these sensors in a hostile environment also causes vulnerabilities and destruction. The lifetime of these sensors is dependable on the application environment and topology, which causes communication disruption and a lack of efficiency, hence the attacks on the Snooping, Sniffing, Compromised Identity, Eavesdropping, etc.

We now turn to an analysis of IoT attacks and security/privacy issues based on the construction of the four layers described above. Figure 5.2 presents the division of the attack categories. This section explains the principal risk factors and mitigation challenges in more detail.

In this chapter, we focus our study on one form of attack, i.e., a Denial-of-Service (DoS) attack [18, 19] on IoT devices and the need for new and capable technologies to protect these devices against such attacks. IoT devices

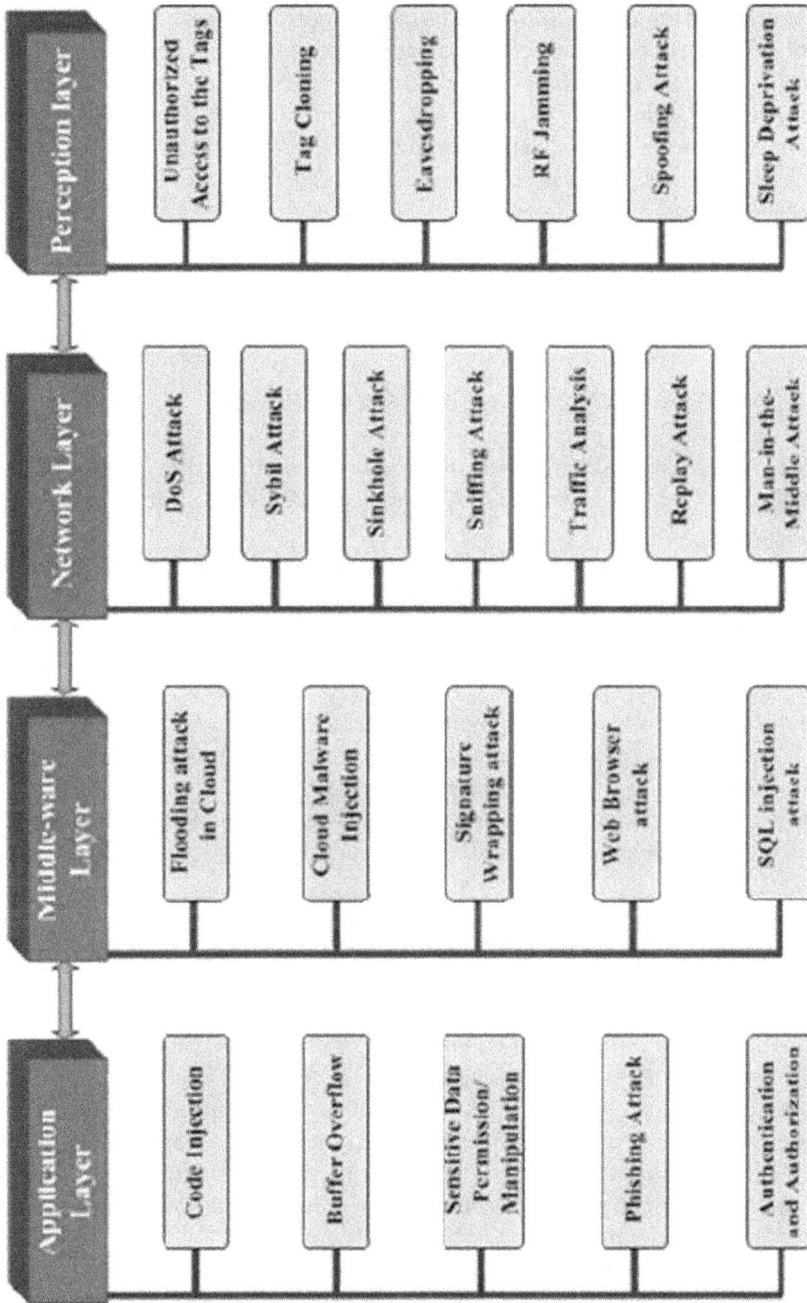

Figure 5.2 Taxonomy of IoT attack at four layers.

serving as botnets target vulnerable devices [20]. For example, the Mirai botnet [21], which affected a significant part of the internet in 2016, affected several countries and employed a large number of geographically distributed IoT devices to orchestrate their attack plan. Such a dispersed contingent of IoT devices becomes challenging to spot and counteract, thereby adding to the seriousness of the attack. A recent report [22] marked that in the months from July to September of 2019, the highest rate of devices hosted by botnets to launch DDoS attacks were from the United States and the Netherlands. This is a problematic situation that demands the attention of the scientific community to explore new strategies to improve security in the IoT infrastructure and reduce the vulnerability of devices to dangerous attacks like DDoS attacks.

5.3 CATEGORIZATION OF SECURITY ISSUES

Table 5.1 represents the mapping of various IoT security threats implied by denial of service (DoS) with the layers they affect, the level of severity, and their respective solutions.

Let us understand the IoT taxonomy based on its deployment infrastructure:

- Low level of security issues: The first level of security deals with security issues in the physical layer, data link layers of communication, and hardware level.
- Medium level of security issues: Intermediate security issues are particularly focused on communication, routing, and session management network at IoT transport layers.
- High level of security issues: High-level security issues are mostly the user or system applications running on IoT.

5.4 SECURITY ISSUES OF THE LOWER LAYERS

Jamming attacks: In this type of attack, the attacker causes communication stoppage by occupying a channel [23], thereby preventing other nodes from using it. This is a particular type of DOS attack. Such attacks work by emitting radio frequency signals without following specific protocols or expected system protocols, hence disrupting networks in wireless devices in IoT. This radio interference [24] affects network performance and may act to and from a communication of official information nodes, leading the system to malfunction.

Sybil and spoofing attacks: Malicious Sybil nodes [25, 26] employ fake identities to launch attacks on wireless networks, eventually degrading the system's performance. Intending to extinguish network resources, Sybil nodes use randomized MAC values impersonating [27, 28, 29] an authentic

Table 5.1 Layer-wise DoS attacks in IoT and mitigation strategies

Attack type	IoT Layer	Level of Attack	Mitigation Strategies
Jamming Attacks	Physical Layer	Low-Level	Evaluating the strength of the signal, delivery ratio of computer packets, encrypting packets with error amending codes, as well as changes to frequencies and locations [21] [22] [23]
Sybil and spoofing attacks	Physical Layer	Low-Level	Measuring signal strength, as well as estimation of the channel [24] [25] [26]
Insecure initialization and configuration	Physical Layer	Low-Level	To set data transfer rates between nodes and introduce artificial noise to the channel [27] [28] [29]
Insecure physical interface	Hardware	Low-Level	Not granting USB or any other hardware-based module or debugging tools access to system software [30]
Duplication attacks due to fragmentation	6LoWPAN adaptation layer and network layer	Intermediate Level	Presentation of Time stamp and nonce protection options against replay attacks and fragment confirmation via hash chains [31] [32]
Sinkhole and wormhole attacks	Network layer	Intermediate Level	Ranking requests via hash chain function, building a level of trust and managing it, analyzing the actions of the nodes. Verification rate through Hash chain function, management of trust level, incorporating IDS to detect irregularities, cryptographic key management, signal strength evaluation, traversal of graph [33] [34] [35] [36] [37] [38]
Session establishment and restarting.	Transport layer	Intermediate level	Long-lived secret key authentication, and key-based symmetric encryption [39] [40] [41]
CoAP security with internet	Application layer and network layer	High-level and intermediate level	TLS/DTLS and HTTP/CoAP mapping, Mirror Proxy (MP) and Resource Directory, TLS-DTLS tunnel, and message filtration using 6LBR [42]
Insecure interfaces	Application layer	High-level	Not allowing weak passwords, check the interface against being at risk of software tools (SQLi and XSS), as well as using HTTPS with firewalls [30]
Insecure software/ firmware	Application layer, transport layer, and network layer	High-level, intermediate level, and low-level	General security updates for software/firmware, usage of file signatures, and encryption with confirmation [30]

device in the physical layer. This results in genuine devices missing network access.

Insecure initialization and configuration: The configuration and initialization of IoT devices [30] at the physical layer without obstructing network services or privacy policies guarantees secure and efficient working of the entire system. Connections at the physical layer also need protection against unauthorized recipients trying to gain access.

Insecure physical interface: Many biological factors pose a severe threat to the efficiency of IoT devices. Imagine the damage that can happen if a harmful individual gains physical access to a task-critical device. Inefficient physical security [31], unauthorized access to software through physical avenues, and testing/debugging tools may compromise the network nodes.

5.5 SECURITY ISSUES OF THE INTERMEDIATE LEVELS

Replay or duplication attacks [32] due to fragmentation: For devices obeying the standards of IEEE 802.15.4, the fragmentation ofIPv6 packets is necessary. When the packet fragment fields are recreated in the 6LoWPAN layer, resources may be depleted [33], a buffer may overflow, and the device may reboot. Duplicate pieces posted by malicious nodes affect packet reconstruction, thereby preventing the processing other genuine packets.

Sinkhole and wormhole attacks: The attacker employs a malicious node to reply [34, 35] to routing requests, bringing packets and contact with hostile agents, and thereby exposing it to mishaps. Network attacks may further exacerbate the performance of 6LoWPAN due to the wormhole attack [36], where a tunnel is built between two nodes so that packets reach from one location to another immediately. This attack has serious consequences, which include eavesdropping [37], breach of privacy, and Denial of Service.

Session establishment and resumption: If a session is hijacked in the transport layer via fake/forged messages, it leaves the system vulnerable to a DoS attack. The attacking node can masquerade [38] as a victim to maintain the session between two nodes. The re-relay of messages by switching sequence numbers may be required.

5.6 SECURITY ISSUES OF THE HIGHER LEVELS

CoAP security [39] with the internet: CoAP, short for Constrained Application Protocol, works at the application layer in the high-level layer and is a web transfer protocol well suited for nodes that run on simple microcontrollers with a limited amount of RAM and ROM and communicate over low-power wireless area networks. It employs DTLS ties with many other ways to protect both ends of the network. Defined in RFC-7252, they follow a scheme that needs encryption to enhance security.

Insecure interfaces: The port recruited to access IoT assets via the web, mobile, or cloud is always susceptible to attacks that threaten data privacy [40].

Insecure software/firmware: IoT devices, when performing updates, go for insecure network protocols, thus exposing themselves to attacks [41] where the attacker can span across the network and remodel it to his desire. He can even infect the other nodes connected via this device's hub.

Middleware security: The middleware [42] layer in IoT serves the purpose of establishing correspondence between various bodies in the IoT paradigm, which, in order to reap the benefit of its service, must be adequately secured.

5.7 MITIGATION STRATEGIES AT VARIOUS LEVELS OF IoT DEVICES

As we have seen, IoT devices have many such avenues that can be plagued by various security issues. Some examples of these avenues are insecure web interfaces, insecure network services, transport lacking encryption, insecure cloud interface, insecure mobile interface, and insecure software/firmware to name a few. Since IoT has become such an indivisible part of our society, threats to its security must be dealt with conscientiously. Here are a few mitigation strategies at different levels of IoT.

Security solutions for the lower level: In wireless sensor networks, the interruptions that lead to message conflicts or channel overload are linked to jamming attacks, as referred to by [23, 24] in the previous section. One of the methods used (as mentioned in these references) to discern whether a jamming attack has occurred is by calculating the effective rate in which a packet is delivered. This design functions by executing statistical tests on the strength of the signal and node address. An alternative to counter jamming attacks involves cryptographic functions and error correction codes. The process progresses by encoding packet codes by breaking it into blocks and leaving pieces of the coded package. Similarly, strategies involving channel filtering and site deceleration have also been proposed to combat jamming attacks. Young et al. [42] proposed a tactic to detect jamming attacks. First the signal strength is measured, then it is utilized to draw out signals that are noise-like. Later these numbers are collated with standard threshold values assigned for attack detection. Channel browsing allows legitimate communication devices to switch channels; this, in turn, causes these devices to alter their address in between the process.

As previously referred to in [27, 28, 29] malicious Sybil nodes impersonate a distinct device by using a forged MAC address which leads to resource exhaustion and denial to grant access to genuine customers. One of the fundamental approaches here is to track Sybil attacks by determining the power of the signal. Detector nodes are positioned in the communication channel to pinpoint the sender's address. Sybil attacks are characterized by a similar sender location, but different sender IDs. Signal strength ratings on

MAC addresses can also be used to detect fraudulent attacks. Another alternative can be the use of Received Signal Strength Indicator (RSSI) for the detection of a Sybil attack. Trust-based models are also very effective approaches to detect Sybil attack in MANET.

In general devices with poor physical security owe their failing to insecure connections to software, firmware and flawed testing/debugging tools. Hardware ports like USB ports or an SD card slot can be easily compromised by granting access to malicious agents. Open Web Application Security Project (OWASP) provides recommendations for an improvement in the physical safety of IoT devices. Testing and debugging tools should be disposed and methods such as Trusted Platform Modules (TPMs) should be brought in to improve physical safety.

Security solutions for the intermediate level: By adding time stamps and nonce options to fragmented packet threats from replay packages can be alleviated. Correlated package sections are connected to the 6LoWPAN adaptation layer. The timestamp option works for unidirectional packets whereas the nonce option works for the bidirectional packets. The 64-bit time stamp on the piece ensures the removal of recurring obsolete headers and diversions from the network. The fact that the header is made only in response to a new application is ensured by the nonce option. Also, to make sure that the order of packets is not disrupted when transferring IPv6 packet fragments domestically via 6LoWPAN.

Riaz et al. [43] propose a security structure with modules to facilitate locating safe neighbour, authenticity confirmation, the generation of a key and the encryption of data. To locate a safe neighbour Elliptic Curve Cryptography (ECC) can be considered. Weekly et al. [44] propose dealing with so-called sinkhole attacks on low-power Lossy networks with a process that combines failover and verification techniques. In rank verification regarding Destination Information Object (DIO) message, one way hash function as well as hash chain function are used.

On network layer Sybil attacks use fake identities to present themselves as many different individual devices, which are termed Sybil nodes [45]. Participation and the distribution of IoT peers to peers (p2p) systems is threatened by this. Additionally, reliable streaming on the network is risked because the defence against Byzantine errors is influenced. For communal interactions, trust building is encouraged to limit a new Sybil ID from emerging. Social graphs are used as a preventive measure to enable the identification of Sybil nodes by either randomly walking through the graph or using certain detection algorithms in the community [46].

Security solutions for the higher level: Brachmann et al. [47] propose a method combining Transport Layer Security(TLS) and Datagram Transport Layer Security (DTLS) to defend CoAP-based low-power, Lossy network (LLN) in touch with the internet. Instances where this method finds most uses are in cases where 6LBR is linked to LLN through the internet to remotely gain access to devices. CoAP and HTTP clients avail the services of

LLN data nodes. To protect connection security at every end of the system, TLS and DTLS mapping is suggested. When the design for mapping computation is effectuated on devices with arrested assets, the system sustains hefty expenses. Granja et al. [48] offer a way to protect the application messages sent via the internet, and aided by various CoAP security options. Some new alternatives to enhance CoAP-related security are SecurityOn, SecurityToken, and SecurityEncap. SecurityOn defends CoAP messages at the application level. SecurityToken helps in authentication and verified access to CoAP services. SecurityEncap [49] employs SecurityOn and principally performs tasks that enable data channeling necessary for authentication to block relay attacks.

The OWASP project presents specific remedies to secure IoT devices. Weak passwords, vulnerable software tools, and not using HTTPS along with the firewalls should be avoided in order to ensure security at a higher level. Additionally, the installed software should continuously be updated only via encrypted channels. Only signed and validated update files must be downloaded from dependable servers.

When securing distributed applications in IoT, encryption and authentication, as established in VIRTUS middleware proposed by Conzon et al. [50], should be considered. While embracing TLS and SASL for data cohesion, authentication, and the encryption of XML stream, middleware employs an event-drive communication approach. Validating the source ensures the secure transmission of data. When web services are enabled with VIRTUS middleware, it provides dependable and manageable IoT interfaces. Otsopack [51] is another remedy to help heterogeneous implementations in a secure manner. A semantic structure acts as a middleware using a semantic format based on Triple Space Computing (TSC).

5.8 DISTRIBUTED DENIAL OF SERVICE (DDoS) ATTACK

Distributed Denial of Service (DDoS) attacks generally result from draining network resources or malfunctioning [52, 53] computer hardware. This means that the resources are unavailable to legitimate users. Consequently, a DDoS attack involves capturing various devices and further using them to send multiple requests [54] to the server/host or exploiting some of their known threats. DDoS attacks can be categorized in three ways [55]:

 i. Application layer attack;
 ii. Resource exhaustion attack; and
 iii. Volume attack

Application layer attack: This includes low-and-slow attacks, GET/POST floods [56], web server attacks, and so on. These attacks exploit the vulnerabilities of Apache, Windows or OpenBSD, and other operating systems. The intention of these attacks is to crash the web server. Often, these attacks are treated as operational errors, as low-volume traffic requires brutality to copy the behavior of legitimate users. Therefore, this attack is ignored even after obtaining multiple visual indications.

Slowloris [57, 58] is the most common attack on the application layer. In this case, partial requests are sent to the web server at specified times to keep several connections open for a more extended period. As a result, the target web server reaches its maximum simultaneous connection pool capacity. This makes the server unavailable to new requests (legitimate) immediately after the start of attacks. To ensure it works, Slowloris uses the minimum bandwidth, which makes it remain undetected for a long.

Resource Exhaustion Attack: These attacks take advantage of commonly used protocols in the network layer to wear out hardware resources like memory, CPU, and storage. Therefore, such attacks are affected not just by the amount of traffic but also by the combination of particular messages.

The typical case of such attacks involves victimizing TCP communication protocol features. TCP SYN Flood, a classic instance of this attack, initiates a three-way TCP handshake. Then by flooding SYN request messages to the target from a fake source address, the attacker indulges the victim in setting up a new link for the hostile user. After setting up the connection, the target waits for an acknowledgment (ACK) from the client to complete the setup link, which is never achieved as the attacker does not send ACK. Eventually, it causes the backlog to exhaust, thereby rendering the option of opening a new connection impossible.

Volumetric Attack: Volumetric Attack [59, 60] sends considerable data traffic to a network to exhaust its bandwidth. Since the bulk of data (100Gpbs) is required, this attack depends more on the volume of data than application layer attacks and resource exhaustion attacks. Volumetric attacks are launched from multiple sources' IP addresses creating data amplification. This is why this type of attack is hard to mitigate manually. The high volume of data requests results in the expansion of response by servers, thus tiring out the target's bandwidth. Domain Name Server (DNS) Amplification and spoofed Network Time Protocol (NTP) are among these types of attack.

5.9 DDOS ATTACK MITIGATION STRATEGIES

The following sections describe effective strategies to reduce DDoS attacks in IoT.

Flow filtering: Flow filtering is the most straightforward method amongst mitigation solutions as it is commonly used on devices that adhere to

OpenFlow [61, 62]. This strategy factors in the header fields of packages arriving at OpenFlow devices to block the flow of devices labeled as malignant. Among the major criteria that are effective during filtering are:

   i. The address of the source
  ii. Address of destination
 iii. Origin port
  iv. Target port
   v. Network layer protocol

Despite being the most efficient option, it risks creating a bottleneck in the communication interface of the controller owing to its dependence on the controller for statistics gathering and packet inspection, aggregating to enhanced delay:

Honey pots: This approach employs systems in a sheltered and secluded atmosphere to mimic a natural target, thus luring malicious agents. All the information collected via honey pots [63] helps enable the research and development in the field of mitigation policies. Despite being a conventional method for attack reduction, it can be used in alignment with SDN to gather data alongside the controller about malicious traffic.

Rate Limiting: In the case of a volumetric attack, a sizeable amount of malicious traffic results in the unavailability [64] of the network due to overcrowding of its communication links. In this method, a controller defines a maximum traffic limit that can be entertained without the system being deemed overcrowded. Once the network tips the edge of overload, it then declines all the incoming packages. This schema is generally used in conjunction with Advanced Package Test Processes.

Moving Target Defense: As the name suggests, moving target [65] defense works by perpetually resetting system characteristics based on a random value set, preventing the system from being susceptible to attacks that would render it unavailable for sizeable periods. Put simply, the system reinvents certain aspects of its configuration to avoid being targeted. Randomization of IP and MAC addresses is generally followed to make getting hold of host and server information difficult from the network while the process is underway. Despite the wide acceptance of MTD techniques to alleviate DDoS attacks, certain limitations cannot be ignored. For instance, the effect regards to cost and performance when installing MDP on a large-scale network.

Traceback: In this method, information from the package title is used to identify the place of launch/the birthplace of the attacker. In general networks, this task is tricky as the network switches cannot recognize the attacker through their falsified source address fields. However, the advantages that can be enjoyed if we regard the holistic view of the

SDN control plane, then mitigation strategies established on traceback are not a bad option.

Request Prioritization: In this method, the flow of the network processes are managed by setting a priority value. Source hosts for every arrival packet at the SDN controller are allocated a default reliability value based on the hosts' traffic history. Suspicious activities may alter these values by reducing them from time to time. The confidentiality value set by the system administrator, if found low for a source, then its flow is denied.

5.10 RESEARCH GAP

DoS/DDoS attacks are prevalent in multiple layers of IoT architecture, and the mitigation strategies that are proposed to date consider a single layer only. Thus, the research gap lies in the following questions:

- Is it worthwhile to address each layer separately in mitigating DoS attacks in IoT?
- What happens when an attacker changes its signature in real time?
- Once a detection strategy is implemented, will it be able to respond to the change in attack signature?
- Is training a system with a fixed value of a data set and testing the system with a real-time data set feasible?
- Is it appropriate to design a generic system capable of detecting DoS attacks by analyzing features from multiple layers in real-time?

To answer the above questions, it is essential to approach a cross-layer attack detection structure in IoT. This is logical because a single-layer detection mechanism cannot address the heterogeneous nature of IoT devices. Instead, grouping together closely related attacks at different layers will be a more efficient approach to understanding the attacks. Thus, designing a system capable of analyzing multiple attack signatures in real time is a central aspect of any future IoT DoS attack detection strategy.

5.11 FUZZY-NEURAL NETWORK-BASED CROSS-LAYER DoS ATTACK DETECTION FRAMEWORK

To create the cross-layer framework for detecting multiple DoS attacks in real time, a Fuzzy-Neural Network-based cross-layer attack detection mechanism is proposed in this chapter (Figure 5.7). As mentioned in the section on the research gap, this system works in two stages and can identify real-time attack signature variations. Since we are explicitly designing a framework for identifying cross-layer attacks, it is evident to start with cross-layer attack features to be selected. The following subsections discuss the feature selection in brief:

a. **Cross-layer attack features selection:**

As a preliminary proof-of-concept of cross-layer D/DoS attack detection framework, we consider two attacks, Resource Exhaustion Attack and Volumetric Attack, as these are very prominent attacks in IoT scenarios. As discussed in the previous section, a Resource Exhaustion Attack is a flooding attack through which the attacker exhausts network resources. TCP SYN Flood Attack (at the Transport Layer) is one type of such an attack. On the other hand, volumetric attacks flood the network by sending a considerable amount of data traffic, thereby exhausting resources. A DIS flooding attack on RPL (Network Layer) routing protocol is one such attack.

The initial and essential step is to analyze attack features for which we have used the NetSim simulator to generate trace files for both attacks (Figures 5.3 and 5.4). One important aspect here is feature selection, which closely defines the attack signature. This is done by using trace files in the NetSim simulator. After features are extracted, each of the attack data sets is created. At this point, data are pre-processed for feature selection and classification. After data pre-processing, these data sets are processed using machine learning algorithms to select specific features contributing to the attack behavior. In the case of TCP SYN Flood attacks, these features are shown in Figures 5.5 and 5.6.

Comparing Figures 5.5 and 5.6, we see that number of TCP_SYN packets sent to destination node 4 is almost double in the attack scenario. In addition, these attacker nodes are 2, 6, 8, 9 and 10, which indicates that the parameters source id, destination id, TCP_SYN packet sent, SYN-ACK sent, and Interval contribute to the attack signature.

Similarly, DIS interval, delay, and DIS packet sent are the parameters that can be considered attack parameters for a DIS flooding attack.

b. **Proposed framework:**

Now we proceed to the first stage of the proposed framework. In the first stage, the cross-layer data set is created by generating traces using a simulator (NetSim), and a Neural Network is trained with this data set. Along with the data set, one extra input is added to the NN, which labels the input data as either normal (0) or attack (1). We call this an input data type (IDT). This helps to train the NN more precisely with a specific data category. The output of the NN is the detection of an attack, i.e., normal (0), SYN flood attack (1), and DIS attack (2).

The second stage starts once the training of NN is completed. Here we use the NN for accurate time attack detection. At this point, it is crucial to reconsider the real-time IDT data set. The training data set

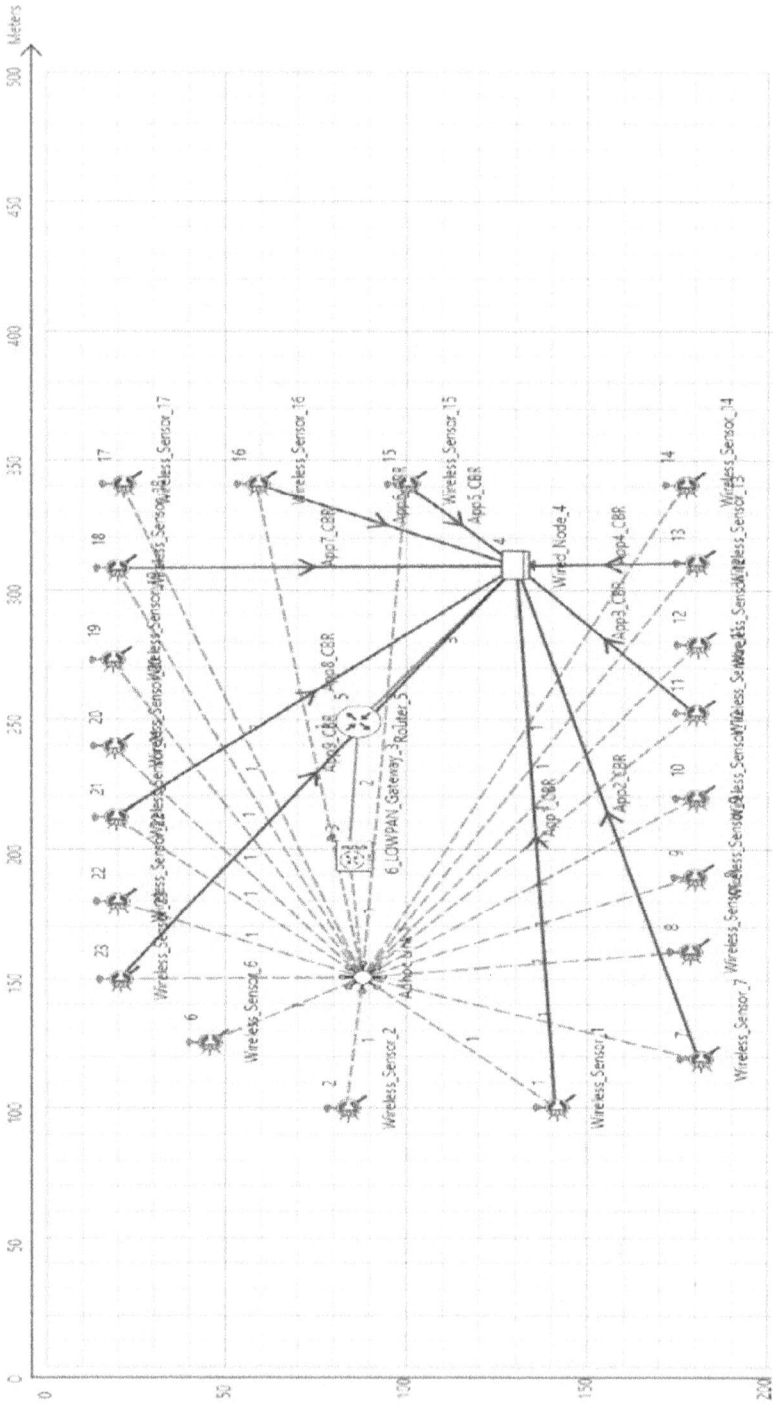

Figure 5.3 TCP SYN Flooding attack.

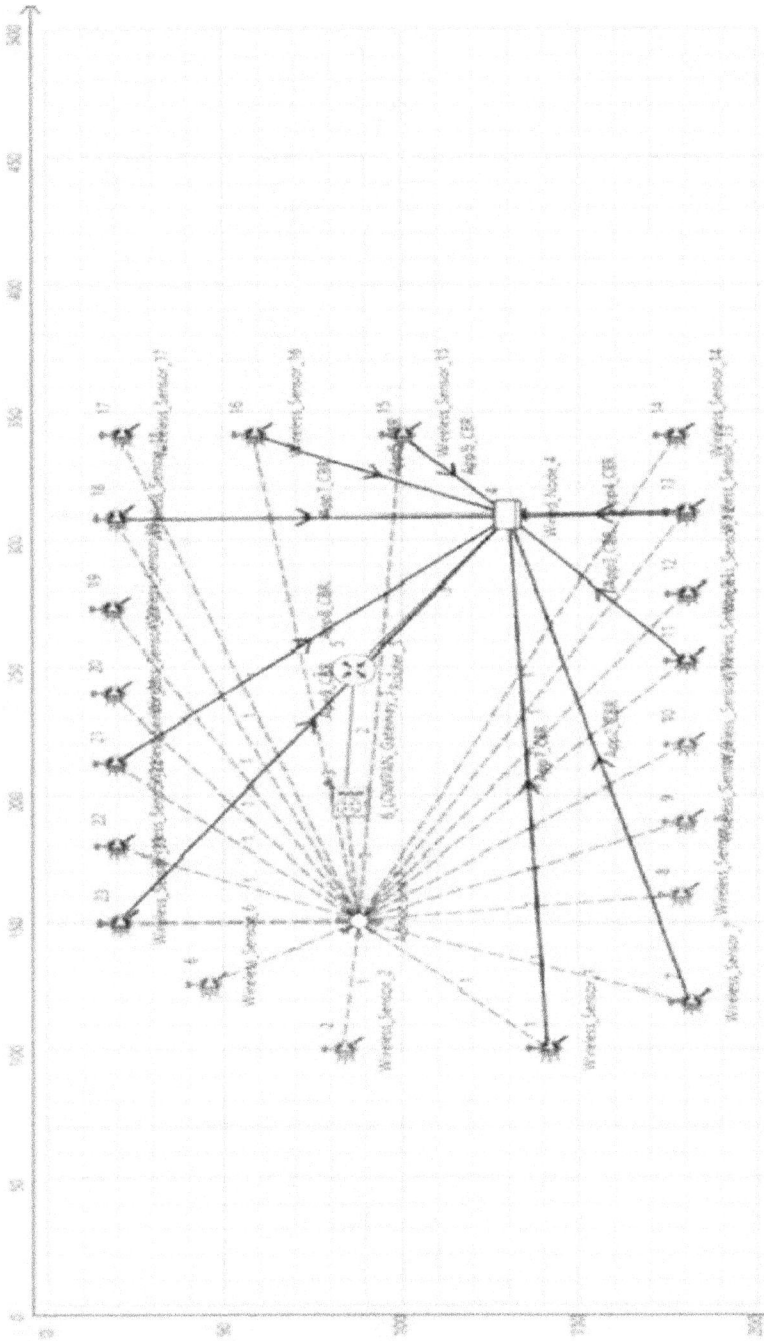

Figure 5.4 DIS Flooding attack.

Count		PACKET_STATUS	DESTINATION_ID	
SOURCE_ID	CONTROL_PACKET_TYPE/APP_NAME		NODE-4	Grand Total
SENSOR-2	TCP_SYN	Successful	3660	3660
	TCP_SYN Total		3660	3660
SENSOR-2 Total			3660	3660
SENSOR-6	TCP_SYN	Successful	3714	3714
	TCP_SYN Total		3714	3714
SENSOR-6 Total			3714	3714
Grand Total			7374	7374

Figure 5.5 TCP SYN Flood attack scenario, Total no. of TCP_SYN packet sent to the destination Node-4 before the attack.

Count		PACKET_STATUS	DESTINATION_ID	
SOURCE_ID	CONTROL_PACKET_TYPE/APP_NAME		NODE-4	Grand Total
SENSOR-10	TCP_SYN	Successful	2694	2694
	TCP_SYN Total		2694	2694
SENSOR-10 Total			2694	2694
SENSOR-2	TCP_SYN	Successful	2780	2780
	TCP_SYN Total		2780	2780
SENSOR-2 Total			2780	2780
SENSOR-6	TCP_SYN	Successful	2604	2604
	TCP_SYN Total		2604	2604
SENSOR-6 Total			2604	2604
SENSOR-8	TCP_SYN	Successful	2602	2602
	TCP_SYN Total		2602	2602
SENSOR-8 Total			2602	2602
SENSOR-9	TCP_SYN	Successful	2634	2634
	TCP_SYN Total		2634	2634
SENSOR-9 Total			2634	2634
Grand Total			13314	13314

Figure 5.6 TCP SYN Flood attack scenario, Total no. of TCP_SYN packet sent to the destination Node-4 after an attack.

is pre-defined, but the real-time data is unpredictable. Thus, the attack data set will have variations compared to the training data set. To reconstruct the IDT, we use the fuzzy rule manager whose output gives us two IDT, viz. normal (0) and attack (1). The fuzzy rule manager works in five steps as discussed below:

Step 1: Create a data set for specific time instances (t1, t2, t3 … tn)
Step 2: For each feature, calculate the average value of all time instances (row average).

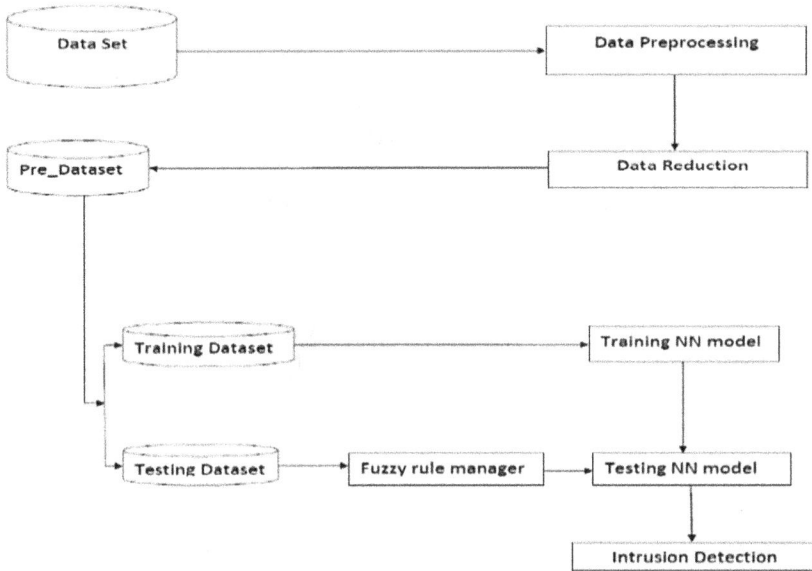

Figure 5.7 Workflow diagram of cross-layer DoS attack detection framework.

Step 3: Calculate the maximum and minimum variation of individual features during the time instances (t1, t2, t3 ... tn)

Step 4: Calculate upper bound = (average feature value + maximum variation) and lower bound = (average feature value – minimum variation) for each feature.

Step 5: If all feature values at any time lie between the upper bound and the lower bound, the feature set represents a normal IDT (0); otherwise, if any of the features' values go above upper bound or below lower bound, it's an attack IDT (1).

The fuzzy rule manager's justification is to analyze each feature's upper and lower bound during a time interval. As long as each component lies between the upper and lower bound, the IDT remains at 0 value, and the fuzzy rule manager goes on checking the next feature. As soon as any part shows a value above or below the upper or lower bound, the process stops at that point and updates IDT as 1. This reduces the processing time of the technique in the case of an attack data set. The value of n is an integer that helps to increase the detection accuracy.

Once the IDT is reconstructed, the input testing data set to the NN is sent in real time, and the output is either expected or any of the two attacks. The importance of IDT is the minimization of a false positive rate due to undetected variation in the data set. This is solved using a fuzzy rule manager, which can enhance the detection accuracy once implemented.

5.12 CONCLUSION

Cross-layer detection framework for DoS attack detection is complex as the exact threshold of attack value is never fixed. With the change of attack signature, the importance of the selected features will change, which may incorrectly interpret an event as an attack, i.e., generation of false positive and false negative, which is revealed from an experimental study. In conclusion, it is observed that cross-layer framework accuracy depends on two significant factors.

- Selection of attack signature as that can vary on an attack basis
- The choice of upper and lower bound for fuzzy rule managers according to the variation of features.

To increase the detection accuracy, the variation in attack signature can be considered dynamically during testing through the NN, which might be effective according to observation. As proposed in this chapter, the fuzzy rule manager can calculate this variation by fixing up the upper and lower bounds of the variation in each feature.

5.13 FUTURE WORK

The heterogeneity of IoT devices drives the implementation of a dynamic attack detection framework to be the most relevant, and possibly the most logical, way of implementing cross-layer DDoS attack detection architecture. Thus, the future aspect is to implement the framework with a real-time data set, increase the detection accuracy by dynamic attack signature selection, and come up with a uniform set of fuzzy rules with upper and lower bounds. Overall, the framework is best to be worked with a signature-based strategy and for a known attack. Still, the framework has to be renewed and taken as future work for unknown environments and attacks requiring an anomaly-based approach.

REFERENCES

[1] M. Lombardi, F. Pascale, and D. Santaniello, "Internet of Things: A General Overview between Architectures", *Information*, 21, p. 1, 2021.
[2] S.K. Jain, and N. Kesswani, "Privacy Threat Model for IoT", In: Nain, N., Vipparthi, S. (Eds.) *4th International Conference on Internet of Things and Connected Technologies (ICIoTCT), Advances in Intelligent Systems and Computing*, vol 1122. Springer, Cham, 2020. https://doi.org/10.1007/978-3-030-39875-0_30
[3] F. Meneghello, M. Calore, D. Zucchetto, M. Polese, and A. Zanella, "IoT: Internet of Threats? A Survey of Practical Security Vulnerabilities in Real IoT Devices", *IEEE Internet of Things Journal*, 6, pp. 1–2, 2019.

[4] K. Chen, S. Zhang, Z. Li, Y. Zhang, Q. Deng, S. Ray, and Y. Jin, "Internet-of-Things Security, and Vulnerabilities: Taxonomy, Challenges", *Hardware and Systems Security*, 2, pp. 97–110, 2018.

[5] C.-K. Wu, *Internet of Things Security, in Advances in Computer Science and Technology*, Springer, Singapore, pp. 1–12, 2021.

[6] S. Checkoway, D. McCoy, B. Kantor, D. Anderson, H. Shacham, S. Savage, K. Koscher, A. Czeskis, F. Roesner, and T. Kohno, "Comprehensive Experimental Analyses of Automotive Attack Surfaces," in *Proceedings of the 20th USENIX Security Symposium*, San Francisco, 2011.

[7] B. B. Gupta and A. Dahiya, "Fundamentals of DDoS Attack: Evolution and Challenges," in *Distributed Denial of Service (DDoS) Attacks*, CRC Press, pp. 1–18, 2021. eBook. ISBN 9781003107354.

[8] https://securelist.com/the-cost-of-launching-a-ddos-attack/77784/

[9] A. Marzano, D. Alexander, O. Fonseca, E. Fazzion, C. Hoepers, K. Steding-Jessen, M. H. P. C. Chaves, Í. Cunha, D. Guedes, and W. Meira, "The Evolution of Bashlite and MiraiIoT Botnets", in *2018 IEEE Symposium on Computers and Communications (ISCC)*, Natal, 2018.

[10] https://www.cloudflare.com/learning/ddos/famous-ddos-attacks/.

[11] https://www.theguardian.com/technology/2016/oct/26/ddos-attack-dyn-mirai-botnet.

[12] M. R. Abdmeziem, D. Tandjaoui, and I. Romdhani, *Architecting the Internet of Things: State of the Art*, in Studies in Systems, Decision, and Control, Springer, Cham, pp. 56–62, 2015.

[13] A. Al-Fuqaha, M. Guizani, M. Mohammadi, M. Aledhari, and M. Ayyash, "Internet of Things: A Survey on Enabling Technologies, Protocols, and Applications", *IEEE Communications Surveys Tutorials*, 17, pp. 2347–2376, 2015.

[14] P. K. Donta, S. N. Srirama, T. Amgoth, and C. S. R. Annavarapu, "Survey on Recent Advances in IoT Application Layer Protocols and Machine Learning Scope for Research Directions", *Digital Communications, and Networks*, 8, 5, pp. 2352–8648, 2021.

[15] A. H. Ngu, M. Gutierrez, V. Metsis, S. Nepal, and Q. Z. Sheng, IoT Middleware: "A Survey on Issues and Enabling Technologies", *IEEE Internet of Things Journal*, 4, pp. 1–20, 2017.

[16] P. Sethi, and S. R. Sarangi, "Internet of Things: Architectures, Protocols, and Applications", *Journal of Electrical and Computer Engineering*, 54, pp. 1–15, 2017.

[17] J. Zhou, Z. Cao, X. Dong, and A. V. Vasilakos, "Security and Privacy for Cloud-Based IoT: Challenges", *IEEE Communications Magazine*, 55, pp. 26–33, 2017.

[18] S. Mishra, and A. Paul, "A Critical Analysis of Attack Detection Schemes in IoT and Open Challenges", *2020 IEEE International Conference on Computing, Power and Communication Technologies (GUCON)*, Greater Noida, India, pp. 57–62, 2020, doi:10.1109/GUCON48875.2020.9231077.

[19] J. Dizdarevic, F. Carpio, A. Lukan, and X. Masip-Bruin, "A Survey of Communication Protocols for Internet of Things and Related Challenges of Fog and Cloud Computing Integration", *ACM Computing Surveys*, 51, 6, pp. 1–29, 2019.

[20] C. Wheelus, and X. Zhu, "IoT Network Security: Threats, Risks, and a Data-Driven Defense Framework", *IoT*, vol 1, pp. 259–282, 2020.

[21] G. Kambourakis, C. Kolias, and A. Stavrou, "The mirai botnet and the iot zombie armies", *Military Communications Conference (MILCOM)*, pp. 267–272. IEEE, 2017.

[22] https://securelist.com/ddos-attacks-in-q3-2021/104796/

[23] W. Xu, W. Trappe, Y. Zhang, and T. Wood, "The feasibility of launching and detecting jamming attacks in wireless networks", in *Proceedings of the 6th ACM*, New York, 2005.

[24] G. Noubir, and G. Lin, "Low-power DoS Attacks in Data Wireless LANs and Countermeasures", *ACM SIGMOBILE Mobile Computing and Communications Review*, Vol. 7, No. 3, pp. 29–30, 2003.

[25] Sinha, S., and A. Paul "Neuro-Fuzzy Based Intrusion Detection System for Wireless Sensor Network". *Wireless Pers Communication*, vol 114, pp. 835–851, https://doi.org/10.1007/s11277-020-07395-y

[26] A. Paul, and S. Sinha, "Performance Analysis of Received Signal Power-Based Sybil Detection in MANET Using Spline Curve", *International Journal of Mobile Network Design and Innovation*, 7, pp. 222–232, 2018, doi:https://doi.org/10.1504/IJMNDI.2017.089304.

[27] S. Sinha, A. Paul, and S. Pal, "The Sybil Attack in Mobile Adhoc Network: Analysis and Detection", in *Proceedings of International Conference on Recent Trends in Communication and Computer Networks-ComNet 2013*, pp. 95–103, 2013.

[28] M. Demirbas, and Y. Song, "An RSSI-Based Scheme for Sybil Attack Detection in Wireless Sensor Networks", *International Symposium on a World of Wireless, Mobile and Multimedia Network*, IEEE, Washington, 2006.

[29] A. Paul, S. Sinha, and S. Pal, "An Efficient Method to Detect Sybil Attack Using Trust Based Model", in *Proceedings of Fourth International Joint Conference on Advances in Engineering and Technology AET-ACS-2013*, pp. 228–237, 2013.

[30] P. Tommaso, L. Brilli, and L. Mucchi, "The Role of Physical Layer Security in IoT: A Novel Perspective", *Information*, 7, 49, 2016.

[31] https://owasp.org/index.php/Top_IoT_Vulnerabilities.

[32] H. Kim, "Protection Against Packet Fragmentation Attacks at 6LoWPAN Adaptation Layer", in *2008 International Conference on Convergence and Hybrid Information Technology*, IEEE Xplore, 2008.

[33] R. Hummer, J. Hiller, H. Wirtz, M. Henze, H. Shafagh, and K. Wehrle, "6LoWPAN Fragmentation Attacks", in *Sixth ACM Conference on Security and Privacy in Wireless and Mobile Networks*, New York, 2013.

[34] K. Weekly, and K. Pister, "Evaluating Sinkhole Defence Techniques in RPL networks", in *20th IEEE International Conference on Network*, Washington, pp. 1–6, 2012.

[35] F. Ahmed, and Y.-B. Ko, "Mmitigation of Black Hole Attacks in Routing Protocol for Low Power and Lossy Network", *Security and Communication Networks*, 9, 5143–5154, 2016.

[36] A. A. Pirzada, and C. McDonald, "Circumventing sinkholes and wormholes", in *International Workshop on Wireless Ad-Hoc Networks*, 2005.

[37] W. Wang, J. Kong, B. Bhargava, and M. Gerla, "Visualisation of Wormholes in Underwater Sensor Networks: A Distributed Approach", *International Journal of Security and Networks (IJSN)*, 3, pp. 10–23, 2008.

[38] N. Park, and N. Kang, "Mutual Authentication Scheme in Secure Internet of Things Technology for Comfortable Lifestyle", *Sensors*, 16, p. 20, 2016.

[39] M. H. Ibrahim, "Octopus: An Edge-Fog Mutual Authentication", *International Journal of Network Security*, 18, pp. 1089–1101, 2015.

[40] S. Raza, D. Trabalza, and T. Voigt, "6LoWPAN Compressed DTLS for CoAP", in *8th International Conference on Distributed Computing in Sensor Systems*, Hangzhou, pp. 287–289, 2012.

[41] H. G. C. Ferreira, R. T. de Sousa, F. E. G. de Deus, and E. D. Canedo, "Proposal of a secure, deployable and transparent middleware for the Internet of Things", *2014 9th Iberian Conference on Information Systems and Technologies (CISTI)*, pp. 1–4, 2014.

[42] M. Young, and R. Boutaba, "Overcoming Adversaries in Sensor Networks: A Survey of Theoretical Models and Algorithmic Approaches for Tolerating Malicious Interference", *IEEE Communication Surveys & Tutorials*, 13, pp. 617–641, 2011.

[43] R. Harkanson, and Y. Kim, "Applications of elliptic curve cryptography: A light introduction to elliptic curves and a survey of their applications", in *Proceedings of the 12th Annual Conference on Cyber and Information Security Research*, Tennessee, 2017.

[44] K. Weekly, and K. Pister, "Evaluating Sinkhole defense techniques in RPL networks", in *20th IEEE International Conference on Network*, Washington, DC, 2012.

[45] H. Yu, M. Kaminsky, P. B. Gibbons, and A. Flaxman, "Sybil Guard: Defending Against Sybil Attacks via Social Networks", *ACM SIGCOMM Computer Communication Review*, 36, pp. 267–278, 2006.

[46] D. Quercia, and S. Hailes, "Sybil attacks against mobile users: friends and foes to the rescue", in *The 29th Conference on Computer Communications (IEEE INFOCOM 2010)*, Piscataway, NJ, USA, pp. 1–5, 2010.

[47] M. Brachmann, S. L. Keoh, O. G. Morton, and S. S. Kumar, "End-to-End Transport Security in the IP-based", in *21st International Conference on Computer Communications and Networks*, Munich, Germany, pp. 1–5, 2012.

[48] J. Granjal, E. Monteiro, and J. S. Silva, "Application-layer security for the WoT: Extending CoAP to Support end-to-end message security for internet-integrated sensing applications", in *International Conference on Wired/Wireless Internet Communication*, St. Petersburg, pp. 140–153, 2013.

[49] M. Brachmann, O. Garcia-Morchon, S.-L. Keoh, and S. S. Kumar, "Security considerations around end-to-end security in the IP-based Internet of Things", In *Workshop on Smart Object Security, in conjunction with IETF83*, Paris, 2012.

[50] D. Conzon, T. Bolognesi, P. Brizzi, A. Lotito, R. Tomasi, and M. A. Spirito, "The VIRTUS Middleware: an XMPP based architecture for secure IoT communications", in *21st International Conference on Computer Communications and Networks*, IEEE, pp. 1–6, 2012.

[51] A. Gómez-Goiri, P. Orduña, J. Diego, and D. López-de-Ipiña, "Otsopack: Lightweight Semantic Framework for Interoperable Ambient", *Computers in Human Behavior*, 30, pp. 460–467, 2013.

[52] C. H. Liu, B. Yang, and T. Liu, "Efficient Naming, Addressing and Profile Services in Internet-of-Things Sensory Environments", *Ad Hoc Networks*, 18, pp. 85–101, 2014.

[53] A. Lohachab, and B. Karambir, "Critical Analysis of DDoS—An Emerging Security Threat over IoT Networks," *Journal of Communications and Information Networks*, 3, 3, pp. 57–78, 2018.

[54] S. T. Zargar, J. Joshi, and D. Tipper, "A Survey of Defence Mechanisms Against Distributed Denial of Service (DDoS) Flooding Attacks", *EEE Communications Surveys & Tutorials*, 15, pp. 2046–2069, 2013.

[55] N. Dayal, P. Maity, S. Srivastava, and R. Khondoker, "Research Trends in Security and DDoS in SDN", *Security and Communication Networks*, 9, pp. 6386–6411, 2016.

[56] R. Vishwakarma, and A. K. Jain, "A Survey of DDoS Attacking Techniques and Defense Mechanisms in the IoT Network", *Telecommunication Systems*, 73, pp. 3–25, 2020.

[57] S. McGregory, "Preparing for the Next DDoS Attack", *Network Security*, 5, pp. 5–6, 2013.

[58] Y. G. Dantas, V. Nigam, and I. E. Fonseca, "A selective defense for application layer DDoS attacks", in *IEEE Joint Intelligence and Security Informatics Conference*, The Hague, The Netherlands, pp. 75–82, 2014.

[59] S. Mansfield-Devine, "The Evolution of DDoS", *Computer Fraud & Security*, 10, pp. 15–20, 2014.

[60] B. Sieklik, R. Macfarlane, and W. J. Buchanan, Evaluation of TFTP DDoS Amplification Attack, *Computers & Security*, 57, pp. 67–92, 2016.

[61] T. Lukaseder, K. Stolzle, S. Kleber, B. Erb, and F. Kargl, "An SDN-based Approach For Defending Against Reflective DDoS Attacks", in *IEEE 43rd Conference on Local Computer Networks*, Chicago, pp. 299–302, 2018.

[62] G. Sun, W. Jiang, Y. Gu, D. Ren, and H. Li, "DDoS Attacks and Flash Event Detection Based on Flow Characteristics in SDN", in *Proceedings of the 2018 15th IEEE International Conference on Advanced Video and Signal Based Surveillance (AVSS)*, Auckland, pp. 1–6, 2018.

[63] X. Luo, Q. Yan, and M. Wang, Huang W, "Using MTD and SDN-based Honeypots to Defend DDoS Attacks in IoT", in *Proceedings of the 2019 Computing, Communications and IoT Applications (ComComAp)*, Shenzhen, China, pp. 392–395, 2019.

[64] S. S. Bhunia, and M. Gurusamy, "Dynamic attack detection and mitigation in IoT using SDN", in *Proceedings of the 2017 27th International Telecommunication Networks and Applications Conference (NAC)*, Melbourne, Australia, pp. 1–6, 2017.

[65] D. Ma, Z. Xu, and D. Lin, "Defending Blind DDoS Attack on SDN Based on Moving Target Defense", In: Tian, J., Srivatsa, M. (Eds.), *International Conference on Security and Privacy in Communication Networks*, Springer International Publishing: Cham, Switzerland, pp. 463–480, 2015.

Chapter 6

Extending the Unified Theory of Acceptance and use of technology model to understand the trainees' acceptance and usage of Internet of Things (IoT) by skill development course

Ananta Narayana
DHSS, MNNIT, Allahabad, India

P. S. Birla
FOE, IGNTU, Amarkantak, India

R. K. Shastri
DHSS, MNNIT, Allahabad, India

CONTENTS

DOI: 10.1201/9781003407300-6

6.1 INTRODUCTION

Technologies related to the Internet of Things system are expanding rapidly. In the world, there are billions of physical IoT devices, all of which are connected to the internet. We can also describe the Internet of Things as a digital world which is embedded with technologies and connects everything through its system and communication networks. The Internet of Things consists of electronic devices and is no longer limited to the original three connected devices: computers, laptops, and smartphones. The Internet of Things is beyond sharing data with further internet-enabled devices. It is important from a technical, social, and economic standpoint to pay attention to the Internet of Things. Consumer goods, durable goods, transportation equipment, industrial and utility components, sensors, and other related items are being integrated with Internet connectivity and potent data analytics capabilities that promise to transform the way we work, live, and play. Impressive predictions have been made as to how the Internet of Things will affect the economy and the number of linked devices, with some predicting that by 2025 there willbe more than 100 billion connected IoT devices and the Internet of Things will contribute some $11 trillion to the global economy. However, there are also significant obstacles raised by these developments that could prevent its potential advantages from being realized. The public's attention has already been piqued by attention-grabbing news concerning the hacking of Internet-connected devices, concerns about surveillance, and privacy anxieties. New issues with regard to policy, law, and development are also emerging, in addition to the ongoing technical difficulties.

The Internet of Things (IoT) is a new paradigm that makes it possible for electrical devices and sensors to communicate with one another over the internet in order to make our lives easier. IoT uses the internet and smart devices to offer creative answers to problems faced by businesses, governments, and both public and private sectors around the world. It is steadily gaining in importance and is now pervasive in our daily lives. IoT, as a whole, is a technological advancement that combines a wide range of smart systems, frameworks, intelligent devices, and sensors. Additionally, it makes use of quantum and nanotechnology to achieve previously unthinkable levels of data storage, sensing, cloud computing, processing speed, and artificial intelligence. To demonstrate the potential effectiveness and applicability of IoT changes, extensive research studies have been conducted and are available in the form of scholarly articles, and press reports, both on the internet and in print form. It could be used as a preliminary backgroundto inform the development of original, inventive concepts while incorporating security, assurance, and interoperability into consideration.

The Internet of Things incorporates a wide range of concepts that are intricate and intertwined from several aspects so as to take the advantage of the Internet of Things and make the technology accessible to the public acrossthe whole of India, the Central Electronics Engineering Research Institute organizes a six-week Internet of Things skill development course to Industry Professionals, Faculties, and B.E./B. Tech students from both the 6th semester onwards and Diploma holders.

For the benefit of society, CSIR-CEERI (CSIR-Central Electronics Engineering Research Institute) is committed to research and development, technology development, and academic assistance (including skill development training/human resources) in the fields of electronics, ICT, and related areas. A prestigious institute has excelled in cutting-edge information and communication technology research and IoT technology is essential for the future of the industry since it keeps the services linked and operating effectively.

The goal of the skill development course is to teach participants how to design and program an Internet of Things (IoT)-based system utilizing real hardware (a Raspberry Pi-3 and an MSP430 board) and the Linux operating system with server connectivity. Connecting IoT devices, such as edge and gateway devices, to the server/cloud, as well as troubleshooting using the Raspberry Pi-3 and MSP430 board and expected job roles include engineers of embedded systems, technicians of embedded systems, and engineers of field applications who can troubleshoot IoT-based electronic systems and products and the creation of small electrical devices based on Internet of Things (IoT) applications, the creation of Internet of Things (IoT)-based solutions for smart cities, intelligent transportation and building systems, environment monitoring and control systems through entrepreneurship.

Course highlights were:

- Expert instruction that is conceptually focused and in-depth, including pertinent updates from associated research teams.
- Training that maximized competency development was made possible by scientific study.
- Expert faculty, lab instructors, and highly skilled trainers with relevant research & development and academic experience.
- Training outcome and competency development is proposed to fulfil criteria specific to industry and academia.

Learning results include:

- Distinguishing between IoT hype and reality
- Raspberry Pi-3 with MSP430 plus CC2520 and Arduino Mega 2560 programming
- Peripheral Raspberry Pi-3 plus MSP430 interface
- Embedded systems based on the Raspberry Pi-3 and MSP430
- Programming methods using LAMP
- Examine networking technologies to be used in Internet of Things applications.

- Analyze the interconnections between IoT, cloud services, and software agents.
- Use practical methods to develop IoT-based projects.

Despite the fact that the idea of using computers, sensors, and networks to monitor and manage items has been around for some time, the recent convergence of important technological advancements and commercial trends is ushering in a new era for the "Internet of Things." IoT promises to introduce a revolutionary, globally networked "smart" future, where connections between people and their environment and between products and their environment will become increasingly intricate. People's perceptions of what it means to be "online" may change significantly if the Internet of Things becomes a common array of connected devices.

A variety of obstacles, including those relating to security, privacy, interoperability and standards, legal, regulatory, and rights issues, as well as the inclusion of emerging economies, could stand in the way of this goal, despite the fact that the potential repercussions are considerable. There are many different stakeholders involved in the Internet of Things, and there are many complicated and changing technological, social, and policy factors. The Internet of Things is now present in our everyday lives; therefore, it's important to manage its problems, reap its benefits, and lower its hazards.

IoT is an important aspect in the development of the so-called "Internet Society" because it represents a developing component of how individuals and organizations are likely to engage with and integrate network connectivity into their personal, social, and professional lives. If we see the development of a divisive discussion that contrasts the potential advantages of IoT with its potential problems, solutions to maximizing the benefits and avoiding the risks of IoT are unlikely to be identified. Instead, to choose the most efficient course of action, educated involvement, debate, and collaboration among a variety of stakeholders are required.

In this study we make a critical assessment and incorporation of eight models of technology from the course related to social psychology to the same degree as social cognitive theory, theory of reasoned action, innovation diffusion theory, theory of planned behavior, technology acceptance models, the model of perceived credibility utilization, a hybrid model combining constructs from technology acceptance model, the theory of planned behavior and the motivational model have resulted in the advanced technology model, Unified theory of acceptance and use of technology by (Venkatesh et al., 2003) plus these multiplicities of illustrious models (Yi et al., 2006) along with theories are used to explicate this relationship involving trainees' attitude, perception and behavior intention (Yousafzai, 2012) regarding the usage of technology (Venkatesh et al., 2012).

The Unified Theory of Acceptance and Use of Technology model is both a predictive tool of adoption and behavior and a valid research instrument (Al-Qeisi, 2009). The UTAUT model turns out to be amongst the most extensively used models (Abu-Shanab et al., 2010) because of its parsimony (Foon & Fah, 2011), its usefulness (Venkatesh & Zhang, 2010) and the cluster of antecedents (Yuen et al., 2010; Tarhini et al., 2015). Furthermore, this model is verified to be advanced from the existing technology acceptance model (Venkatesh et al., 2003).

The following are explicit research objectives that can provide insights into the most significant aspects either hindering or facilitating the acceptance and usage of the Internet of Things by trainees from the skill development course.

- To examine which aspect to be precise performance expectancy, effort expectancy, social influence, facilitating conditions, computer self-efficacy and relative advantage affect trainees' behavior intention in the usage of the Internet of Things.
- To identify the characteristics and effectiveness of the relationship between the constructs and elucidate whichever construct influences the decision to a greater extent to use the Internet of Things.
- To analyze the incorporation of two constructs, computer self-efficacy and relative advantage, with a unified theory of acceptance and use of technology model in providing a robust hypothetical basis in favor of investigating the acceptance of the Internet of Things.
- To drive relevant suggestions and recommendations to research institutes in order to ensure a successful training of Internet of Things skill development course.

In general, this chapter will prove to be useful for the government to persuade for adopting and accepting the Internet of Things and for research institutes to provide more robust internet of things skill development courses that adhere better to the market demands for professionals in India. The next section gives atheoretical background which helps in developing a hypothetical research model that influence the adoption of the Internet of Things as the implementation of effective and efficient skill is required for the adoption of innovative technologies (Finis Welch, 1970).

6.2 THEORETICAL BACKGROUND AND HYPOTHESIS

The Internet of Things (IoT) was first introduced by Kevin Ashton in 1999. At the time he defined the IoT as a network of radio-frequency identification (RFID)-enabled technology that is uniquely identifiable and interconnected (Pretz, 2013). The Internet of Things (IoT) was generally described as a "dynamic global network infrastructure with self-configuring capabilities based on standards and interoperable communication protocols; physical and virtual 'things' in an IoT have identities and attributes and are capable of using intelligent interfaces and being integrated as an information

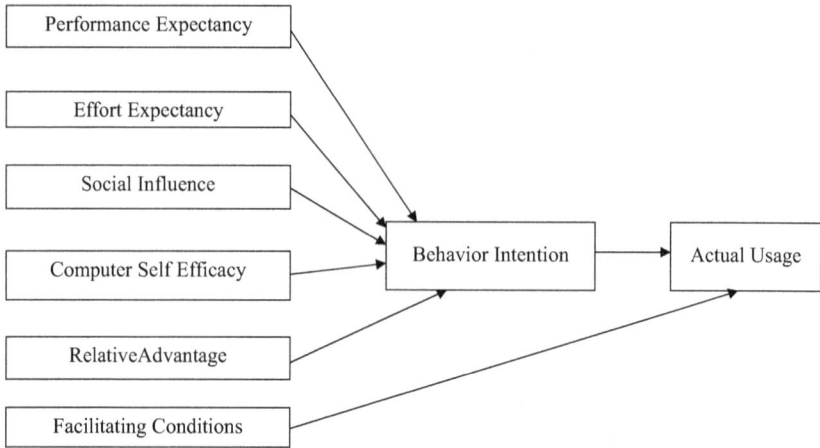

Figure 6.1 Hypothetical research model (adapted from Venkatesh et al., 2003).

network" (Kirtsis, 2011). In essence, the Internet of Things (IoT) can be thought of as a superset of interconnecting objects that can be individually identified using current near-field communication (NFC) technologies. The terms "Internet" and "Things" refer to a global network of interconnected devices that uses networking, information processing, and sensory technologies. This network may represent the future of Information and Communications Technology (ICT) (Kranenburg & Anzelmo, 2011). Despite disagreements over its definition, the Internet of Things has been widely debated and relevant technologies have been created rapidly by a number of institutions. In particular, intelligent sensing and wireless communication techniques have entered the IoT, opening up new problems and research frontiers. The International Telecommunication Union (ITU) explored IoT implications, growing difficulties, emerging markets, and enabling technologies (ITU, 2013).

This section shows the hypothetical research model in Figure 6.1, which is proposed on the basis of the theoretical background and provides a comprehensive discussion of presented constructs

6.3 PERFORMANCE EXPECTANCY

Yu (2012) states that performance expectancy shows a significant effect on the trainees' behavior intention in terms of their actual usage of the Internet of Things. Performance Expectancy measures the degree to which a trainee accepts the Internet of Things as the key to keeping the services interconnected and working effectively (Al-Somali et al., 2009). Correspondingly, Lee (2009) determined performance expectancy as an important factor in understanding the usage of the Internet of Things. With regard to the perspective of the current research study, especially predictable to facilitate the trainees by providing an Internet of Things skill development course to be valuable, so that they can possibly use and accept the Internet of Things. Therefore, the researcher suggests the below hypothesis:

H1. Performance Expectancy significantly affects trainees' Behavior Intention on the usage of the Internet of Things

6.4 EFFORT EXPECTANCY

Venkatesh et al. (2003) define effort expectancy as the degree of effort in which trainees are affiliated with the actual usage of technology. In accordance with Venkatesh's model, effort expectancy positively modifies the behavior intention regarding technology usage. Trainees are more likely to adopt if it does not require much effort and finds the Internet of Things easy to use (Yu, 2012). Yoon and Steege (2013) found that effort expectancy has an effect and proves the strongest predictor of behavior intention. As per, Martins et al. (2014) effort expectancy shows a positive significant effect on behavior intention. Consequently, Zhou et al. (2010) fail to maintain the positive association between effort expectancy and behavior intention. The present research study predicts that if the trainees find the Internet of Things skill development course relevant and easy to use, then they possibly use and adopt it. Contradictorily, if the trainees find the Internet of Things irrelevant and not easy to use, then the trainees will show less interest in adopting it. Hence, the researcher hypothesizes the subsequent hypothesis:

H2. Effort Expectancy will significantly affect trainees' Behavior Intention in the usage of the Internet of Things

6.5 SOCIAL INFLUENCE

Ajzen (1991) determined Social Influence as a particular belief of an individual who is important to the other individual and does not assume the behavior in their perception like social norms mentioned in renowned theories theory of reasoned action, technology acceptance model 2, innovation diffusion theory and theory of planned behavior (Amin, 2009). To put it another way, social influence indicates the societal pressure that comes from the surroundings and uncontrollable factors and can influence the behavior and perception of individuals in certain actions, such as the judgment of superiors, friends, or relatives (Tarhini et al., 2013b; Tarhini & Liu, 2014). While Davis (1989) excluded the social influence factor due to theoretical and measurement tribulations from the original technology acceptance model and was added further in the technology acceptance model 2 due to its significance in elucidating the external influence of other people on the behavior of self (Venkatesh & Zhang, 2010). Social influence persuades technology acceptance (Venkatesh & Zhang, 2010). Social influences refers to the degree to which the use of the Internet of Things boosts a trainee's social status (Moore & Benbasat, 1991). As per Galan et al. (2013), trainees believe that adopting the Internet of Things will improve their performance for a particular social group (Im & Kang, 2011). Most of the experimental research in information and communication technology (Yousafzai, 2012) found social influence to be a significant precursor of behavior intention (Venkatesh et al., 2003) with regard to the Internet of Things (Kesharwani & Bisht, 2012). Mentioned researches indicate a positive significant relationship which is empirically allying linking social influence and behavior intention of trainees' in adopting the Internet of Things.

H3. Social Influence indicates significant influence on trainees' behavior intention in using the Internet of Things

6.6 COMPUTER SELF-EFFICACY

The word self-efficacy is introduced by Bandura (1986). In his social cognitive theory he explained self-efficacy as individuals' belief in their capabilities to accomplish a task confidently within a specified period. Inspired by the idea of the conceptualization of self-efficacy, Compeau and Higgins (1995a) derived the term computer self-efficacy and further explained it as an individual's belief in their capabilities to complete a computer-related assignment confidently within a specified period (Compeau& Higgins, 1995b). In the general computing realm, Marakas, Yi, & Johnson (1998) also illustrated computer self-efficacy as individuals' ability in performing computer-specific tasks. The person who possesses higher computer self-efficacy to own higher confidence in attaining computer-associated tasks successfully. Fagan et al. (2004) specified how computer self-efficacy is positively inclined by motivating teams, colleagues, and management. Computer self-efficacy also controls feelings of anxiety (Tung & Chang, 2008) as individuals who possess a high degree of computer self-efficacy to feel less anxiety when compared with a lower level of self-efficacy. Previous literature found that computer self-efficacy influences individual behavior in adapting to information technology, information, and communication technology; internet banking, etc. so considering these studies researcher inculcates that computer self-efficacy can also be taken as a construct in influencing the behavior intention of individuals in adopting the Internet of Things. Hence, we can formulate the hypothesis as below:

H4. Computer Self-Efficacy will positively affect trainees' behavior intention towards usage of the Internet of Things.

6.7 RELATIVE ADVANTAGE

Relative advantage is a degree that demonstrates the advantage of new technology over existing one and leads to an increase in the adoption rate (Ekong et al., 2012). Rogers (1995) defined relative advantage as "the degree to which an innovation is perceived as being better than the idea it supersedes." As per Tornatzky and Klein (2012), relative advantage shows a significant positive relation with the adoption of innovative technology. Relative advantage shows considerable advantage in the adoption of the Internet of Things (Weber, 2010). Tu (2018) further stated that, as compared to existing offerings, new innovation and technology give considerable benefits. We have come to know from existing literature that relative advantage can prove to be a positive antecedent in influencing behavior intention for technology adoption. Therefore, we articulate the hypothesis as

H5. Relative advantage indicates a positive significant influence on trainees' behavior intention toward usage of the Internet of Things

6.8 FACILITATING CONDITIONS

Venkatesh et al. (2003) elucidate facilitating condition as the degree to which people think about the departmental and technological infrastructure that plays a vital role in technology usage. Undoubtedly, Zhou et al. (2010) indicated that the actual usage of the Internet of Things needed requisite resources and technological infrastructure. Further, certain amenities are never provided generally at any cost with regard to trainees. There are many possibilities for organizations to promote the Internet of Things by eliminating barriers to usage and persuading adoption. Trainees perceive the Internet of Things to be a crucial service and required the use of the most recent technology. Facilitating conditions can be measured on the basis of the trainees' perception that will they can be able to use the requisite resources and services related to the Internet of Things. Consequently, it is predictable that the necessary requisite resources will effectuate the trainees to use and accept the Internet of Things. Hence, based on the above theoretical background our next hypothesis would be as stated below

H6. Facilitating conditions indicate a positive influence on trainees' actual usage of the Internet of Things

6.9 BEHAVIOR INTENTION AND ACTUAL USAGE

Venkatesh et al. (2003, 2012) elucidated behavior intention to be a direct precursor of usage behavior and provide an indicator (Tarhini et al., 2013a) in relation to measuring individuals' ability in adopting a particular behavior

(Venkatesh & Zhang, 2010). Actual behavior is the clear, obvious reaction in a particular situation regarding a particular target (Ajzen, 1991). There is a substantial indication of the significant effect of behavior intention (Im & Kang, 2011) on actual usage in studies related to technology adoption (Ma et al., 2010). In recent times, it has extended to the perspective of the Internet of Things (Yu, 2012). As a consequence, the researcher postulates the below hypothesis:

> H7. Trainees' behavior intention indicates a positive significant influence on their actual usage of the Internet of Things.

6.10 RESEARCH METHODOLOGY

As per previous research studies on the UTAUT Model and the Internet of Things, a quantitative research approach is immersed by the researchers to test the proposed hypothetical model. A structured questionnaire consisting of 38 questions is employed for data collection through convenient sampling from the trainees of the Internet of Things skill development course organized by CSIR-Central Electronics Engineering Research Institute. Trainees were requested to fill out the presented questionnaire on the basis of his/her judgment and attitude about the acceptance and use of the Internet of Things. Each trainee took 15 minutes to fill out the complete questionnaire. The researcher distributed a total of 700 questionnaires to the trainees;of these some 522 were returned to the researcher,meaning a response rate of some 70%. After the researcher review 14 questionnaires were found to be invalid, meaning that 508 responses were selected for final analysis. Table 6.1 shows the beneficiaries' demographic summary; with regard to gender, 63% of respondents are male and 37% are female, their age ranges from 18 to 60 years. With regard to education, 8% of them had secondary education, 34% did a diploma and 58% of them are bachelor's and above, of which 52% of total respondents have experience in using the Internet of Things, 36% have experience of using the Internet of Things to some extent, and 12% have no experience of usage of the Internet of Things.

6.11 MEASUREMENT

The scales were adopted from existing literature associated with the unified theory of acceptance and use of technology model pertaining to the current investigation (Foon& Fah, 2011) and prior empirical studies were taken related to computer self-efficacy and relative advantage (Im & Kang, 2011) to protect the validity and reliability of presented items. Purposely, performance expectancy, and effort expectancy were measured by means

Table 6.1 Beneficiaries demographic summary

Categorization	Freque ncy	Percent
Gender		
Male	321	63.10
Female	187	36.81
Age		
18 to 25	217	42.71
26 to 35	134	26.37
36 to 45	105	20.66
Above 46	52	10.26
Educational level		
Bachelors and above	296	58.26
Diploma	170	33.46
Secondary education	42	8.28
Experience in usage of Internet of Things		
Experienced	264	51.96
Some experience	185	36.43
No Experience	59	11.61

of 5 items, while behavior intention, social influence, and facilitating conditions were measured by 4 items. 2 items were adapted from the work of Venkatesh et al. (2003, 2012). Additionally, computer self-efficacy and relative advantage were measured using 5 items and the scales were adopted from the works of Murphy et al. (1989) and Moore& Benbasat (1991) respectively. In the direction of measuring particular factors of this proposed hypothetical research model, the researcher used a seven-point Likert scale, which ranges from 1 to 7; that is, from strongly disagree to strongly agree, and a nominal scale is used to measure the demographic figures concerning the beneficiaries likewise gender, age, experience, and education. The researcher is unable to take the actual use of the Internet of Things through the trainee's log file as it was not feasible so they measured actual usage using a self-structured questionnaire. The foremost question measured how frequently the trainees are using the Internet of Things while the second question measured the average usage of the Internet of Things by the trainees. The questionnaire is well structured in English, and the items are adapted from the perceptive UTAUT model.Although before proceeding study further, the researchers executed a pilot test of 50 respondents who were randomly chosen for revising and modifying the questionnaire items and establishing content validity and reliability. Several items are examined and adjusted through the results of pilot testing.

6.12 DATA ANALYSIS AND RESEARCH FINDINGS

6.12.1 Descriptive analysis

Table 6.2 presents descriptive statistics which affect all constructs in the hypothetical research model and each mean was > 4.49, indicating that the majority of the respondents respond positively to the structures measured in the study. Furthermore, Cronbach's α score pointed toward the strong internal reliability of all constructs.

6.12.2 Measurement model

The present study is based on a multiple steps approach which examines the relationship between factors of a structural model from the study (Anderson & Gerbing, 1988). The researcher analyzed the measurement model to check whether the instrument is reliable and valid for testing research hypotheses in the proposed hypothetical model (Arbuckle, 2009). Hence, to contemplate the measurement model fit, researchers have to initially analyze the confirmatory factor on the basis of AMOS 20.0 and subsequently calculate the validity appertaining to the measurement model (Schumacker & Lomax, 2010). To offer an approximate calculationof the model's factors, the present examination accepts the maximum-likelihood method where all of the investigation was carried out on variance and covariance matrices. Conversely, before the analysis it is assumed that it is crucial to test for multicollinearity, and multicollinearity depends on a higher correlation between the variables. Tabachnick and Fidell (2007) illustrated that correlation values in the approximate range of 0.8 or 0.9 are supposed to be highly problematic, whereas a correlation value of 0.7 or higher is supposed to be a cause of study. As per Pallant (2010), correlation value tolerance and variance inflation aspect determined the presence of multicollinearity. There would be no chance of multicollinearity on the assumption that the tolerance value is in excess of 0.10 and the value of the variance inflation factor is below 3.0.

Table 6.2 Measures of central tendency and variability

Constructs	Mean	Standard deviation	Cronbach's α
Performance Expectancy	4.61	1.24	0.912
Effort Expectancy	5.41	1.12	0.909
Social Influence	4.26	1.29	0.836
Computer Self Efficacy	4.49	1.38	0.918
Relative Advantage	4.92	0.99	0.862
Facilitating Conditions	5.13	1.27	0.793
Behavior Intention	4.97	1.39	0.857
Actual Usage	5.24	1.37	0.769

The sample in our study shows the nonexistence of multicollinearity as the variance inflation value cause is below 3.0 and the tolerance value is more than 0.10. Hairetal and Kline (2010) suggested various fit indices in support of evaluating the model's goodness-of-fit. This was determined by χ^2, which is termed as the minimum fit function. The fraction of χ^2 is static to its degree of freedom, which is indicative of an acceptable fit (χ^2/df) (Hu & Bentler, 1999; Carmines & McIver, 1981). As per the recommendation of Hair et al. (2010), there is a particular scale of additionally fit indices those indices consist of "goodness-of-fit indices; parsimony normal fit indices; normal fit indices; root mean square residuals; the root mean square error of approximation; adjusted goodness-of-fit indices; and comparative fit indices". From the initial measurement model, researchers removed a few indicators, Social Influence 3, Facilitating Conditions 4, Computer Self Efficacy 5 as well as relative Advantage 6 to sustain a good fit between the model and data. Table 6.3 clearly lists the actual estimate of the model fit index in the suggested range below, hence the researcher can continue to evaluate and validate the discriminant validity and convergent validity along with reliability for checking the adequacy related to psychometric values pertaining to the measurement model. Convergent validity validates whether all the factors got rejected through its indicators conducive to ensuring unidirectional multiple-item constructs by eliminating unpredictable indicators (Gefen et al., 2000; Bollen, 1989). Discriminant validity evaluates a degree to measure the concepts of distinct statistics while composite reliability, maximum shared squared variance, average variance extracted and average shared squared variance can all assess reliability, convergent validity, and discriminant validity. To establish reliability, composite reliability should be at least 0.6 and preferably above 0.7 and to establish convergent validity the average

Table 6.3 Outline of absolute measurement and structural model fit index

Fit index	Recommended value[a]	Measurement model	Structural model
Minimum it Function χ^2	NS at p is less than 0.05	582	597
Degrees of Freedom	n/a	347	356
The ratio of the χ^2 static to its degree of freedom (χ^2/df)	less than 5 preferable less than 3	2.667	2.669
Goodness-of-Fit	Is greater than .9	0.936	0.938
Adjusted Goodness-of-Fit	Is greater than .8	0.885	0.887
Comparative Fit Index	Is greater than .9	0.945	0.944
Root Mean Square Residuals	Is greater than .10	0.086	0.088
Root Mean Square Error of Approximation	Is greater than .09	0.065	0.065
Normed it Index	Is greater than .9	0.950	0.946

[a] Hu and Bentler (1999) Kline (2010) Hair et al. (2010).

variance extracted should be at least 0.5 (50% of the variance of indicators has to be accounted for by the latent variables) and composite reliability is greater than the average variance exceeds. Whereas discriminant validity is supported if maximum shared square variance is less than average variance extracted (Hair et al., 2010). If the extraction of average variance exceeds the shared squared variance and the maximum shared square variance,the resulting discriminant validity is supported. Table 6.4 shows constructs of composite reliability, ranging from 0.76 to 0.91 and exceeding the threshold value of 0.7; at the same time, average variance extracted ranging from 0.53 to 0.77, all of which is above 0.5, resulted in adequate internal consistencies and also supported convergent validity. Furthermore, the maximum shared square variance for all constructs is less than the average variance extracted, resulting in sufficient discriminant value which is illustrated on the basis of a higher square root value of extraction of the average variance than its correlation value. Based on the examination of the measurement model, it is concluded that each variable represents a reliable and valid factor. Therefore the next step is to assesses the structural model in order to test the research model and examine the hypotheses. Consequently, the final measure is headed toward a structural model for testing the hypothetical research model.

6.12.3 Structural model

Owing to the measurement model's criteria, researchers evaluate the goodness-of-fit for measuring the proposed hypothetical research model, and, according to Table 6.3, the results of both the structural and the measurement models were found to be the same, which again showed good fit data. Hence, we continue to study the hypothesized relations of the factors of the proposed hypothetical model. As we see in Table 6.5, all hypotheses except H_2 were supported consequently due to the path coefficient. Notably, performance expectancy, effort expectancy, social influence, and computer self-efficacy all indicated a significant positive effect toward behavior intention about the usage of the Internet of Things and performance expectancy shows the maximum influence on the correlation with behavior intention. The presented results supported Hypothesis-1, Hypothesis-3, Hypothesis-4 and Hypothesis-5 unpredictably; the path coefficient from effort expectancy to behavior intention was not significant. As a result, this study failed to find support for Hypothesis-2. Performance expectancy, social influence, computer self-efficacy, and relative advantage to account for 63% of the variance of behavior intention than other added factors. Contrary to the effect of the statistical analysis of the presented proposed model explained in the context of facilitating condition and behavior intention indicated significantly influence the actual usage of the system and jointly accounted for 66% of the variance in actual usage, with behavior intention contributing the most compared to facilitating conditions, consequently, Hypothesis-6

Table 6.4 Construct reliability, convergent and discriminant validity

Constructs	Composite Reliability	Average Variance Extracted	Maximum Shared Square Variance	Average Shared Square Variance	Computer Self-Efficacy	Performance Expectancy	Effort Expectancy	Social Influence	Relative Advantage	Facilitating Conditions	Actual Usage	Behavior Intention
Computer Self-Efficacy	0.924	0.783	0.47	0.276	0.889							
Performance Expectancy	0.916	0.674	0.327	0.229	0.373	0.87						
Effort Expectancy	0.924	0.759	0.339	0.277	0.579	0.574	0.879					
Social Influence	0.846	0.542	0.275	0.189	0.497	0.463	0.386	0.724				
Relative Advantage	0.873	0.624	0.487	0.286	0.481	0.526	0.532	0.436	0.788			
Facilitating Conditions	0.893	0.753	0.433	0.292	0.657	0.364	0.586	0.468	0.522	0.877		
Actual Usage	0.774	0.542	0.446	0.242	0.523	0.394	0.389	0.324	0.438	0.594	0.723	
Behavior Intention	0.874	0.717	0.487	0.347	0.527	0.576	0.584	0.538	0.584	0.537	0.574	0.854

Table 6.5 Consequence of path coefficients

	Path	Path co-efficient	Results
Hypothesis-1	Performance Expectancy→ Behavior Intention	0.271**	Supported
Hypothesis-2	Effort Expectancy → Behavior Intention	0.084	Not supported
Hypothesis-3	Social Influence→ Behavior Intention	0.223**	Supported
Hypothesis-4	Computer Self-efficacy → Behavior Intention	0.143*	Supported
Hypothesis-5	Relative Advantage → Behavior Intention	0.228**	Supported
Hypothesis-6	Facilitating Conditions → Actual Usage	0.186**	Supported
Hypothesis-7	Behavior Intention → Actual Usage	0.464***	Supported

* p is less than 0.05;
** p is less than 0.01;
*** p is less than 0.001; NS p is greater than 0.01

and Hypothesis-7 were supported. The proposed hypothetical research model of the present study elucidates more variances of behavior intention and actual usage of the Internet of Things.

6.13 DISCUSSION AND CONCLUSION

A new paradigm called the Internet of Things (IoT) has transformed traditional living into a high-tech lifestyle. The changes brought about by IoT include smart cities, smart homes, pollution control, energy conservation, smart transportation, and smart industries. Many important research projects and investigations have been carried out in an effort to advance technology through IoT. Globally, academics and developers are interested in recent IoT breakthroughs. Researchers and IoT developers are collaborating to expand the technology to the fullest possible extent and to produce improvements in the system as a whole. The present research study is determined by extending the unified theory of acceptance and usage of technology by integrating constructs such as computer self-efficacy and relative advantage which study the indicators affecting trainees' behavior intention in terms of usingthe Internet of Things. The findings of the study support the efficiency and aptitude of Venkatesh's technology model,both theoretically and empirically, to be a functional theoretical model for giving insight into trainees' acceptance of the Internet of Things. Path coefficients of the proposed hypothetical model indicated statistical significance between effort expectancy and behavior intention. In particular, the above outcomes indicated that the trainees' intention in terms of the acceptance and usage of the system of the Internet of Things could be significantly affected by performance expectancy, social influence, computer self-efficacy, and relative advantage. By contrast, effort expectancy was not found to be particularly influential with regard to behavior intention in using the Internet of Things. The results of this study also indicated that facilitating conditions and behavior intention, respectively, proved a significant antecedent of the actual use of the Internet of Things. Taken as a whole, the proposed research model attained acceptable fit and account for 63% and 66% of the variances, respectively, for behavior intention and actual use i.e., figures normally higher than the novel model unified theory of acceptance and use of technology. While taking performance expectancy into consideration it proved the strongest antecedent that reflects the trainees' perception for using the Internet of Things in a better way. For that reason, skill development course instructors need to enhance the effectiveness of the Internet of Things system. The results of the study specify that social influence has a positive significant effect on behavior intention, suggesting that trainees become influenced by peer pressure. Thus, skill development course instructors should understand that social influence has an importanteffecton trainees' intention to accept the Internet of Things. The outcome of the present research confirms that facilitating conditions played a significant role in the actual usage of the Internet of Things system, so for that reason there is a need to invest more on information and communication technology infrastructure. For instance, after the training period skill development course instructors should also provide facilities, such as full infrastructure training centers to increase the skills in using the Internet of Things so that the interest of trainees in adopting skills might increase. Whereas effort expectancy unexpectedly shows an insignificant predictor of trainees' intention to usethe Internet of Things as the trainees give priority to the perceived usefulness of the system rather than its ease of use. There is thus a need to design a more user-friendly system which encourages trainees to adopt the system. Consequently, this study proved the importance of indicating a significant positive influence between computer self-efficacy and behavior intention to derive interest in adopting the Internet of Things since computer self-efficacy is already a proven factor in enhancing performance when dealing with computers (Cocorada, 2014). Prior command of computer self-efficacy is thus needed before any such training coursesince it leads to a lower level of anxiety during training and perceived high levels of acknowledged usefulness towardthe adoption of technology (Downey & Kher, 2015). Accordingly, empirical verification of the current research demonstrates that relative advantage has proven a significant factor in deriving behavior intention in using the Internet of Things because it leads to an improved use of resources, efficient operation management, cost-effective operation, and improved performance. Therefore, government and research institutes need to understand that the most important factor concerns trainees' perception to adopt the Internet of Things as the rate of acceptance of the technology is directly proportional to its usage. The greater the usage, the greater will be its acceptance. Expertise in the Internet of things is required for both upskilling and career advancement. Hence, research institutes should emphasize on providing more robust Internet of Things skill development courses that will better addressthe market demands for professionals in India.

REFERENCES

E. Abu-Shanab, J. M. Pearson, and A. J. Setterstrom, "Internet banking and customers' acceptance in Jordan: The unified model's perspective," *Communications of the Association for Information Systems*, vol. 26(1), pp. 23, 2010.

I. Ajzen, "The theory of planned behavior," *Organizational Behavior and Human Decision Processes*, vol. 50(2), pp. 179–211, 1991.

K. I. Al-Qeisi, "Analysing the use of UTAUT model in explaining an online behavior: internet banking adoption," unpublished doctoral dissertation, Brunel Business School, Brunel University, London, 2009.

S. A. Al-Somali, R. Gholami, and B. Clegg, "An investigation into the acceptance of online banking in Saudi Arabia," *Technovation*, vol. 29(2), pp. 130–141, 2009.

H. Amin, "An analysis of online banking usage intentions: An extension of the technology acceptance model," *International Journal of Business and Society*, vol. 10(1), pp. 27–40, 2009.

C. Anderson, and D. W. Gerbing, "Structural equation modeling in practice: Areview and recommended two-step approach," *Psychological Bulletin*, vol. 103(3), pp. 411–423, 1988.

J. Arbuckle, *AMOS 18 User's Guide*, SPSS Incorporated, Armonk, NY, 2009.

A. Bandura, *Social foundations of thought and action: A Social Cognitive Theory*,Prentice Hall, Upper Sadle River, NJ, 1986.

K. A. Bollen, *Structural equations with latent variables*,John Wiley and Sons, Oxford, 1989.

G. Carmines, and J. P. McIver, *Analyzing models with unobserved variables: Analysis of covariance structures*,SAGE, Thousand Oaks, CA, 1981.

S. Cocorada, "Computer anxiety, computer self-efficacy and demographic variables," the *International Scientific Conference eLearning and Software for Education*, vol. 1, Carol I, National Defence University, 2014.

D. R. Compeau, and C. A. Higgins, "Computer self-efficacy: Development of a measure and initial test," *MIS Quarterly*, vol. 19(2), pp. 189–211, 1995a.

D. R. Compeau, and C. A. Higgins, "Application of social cognitive theory to training for computer skills," *Information Systems Research*, vol. 6(2), pp. 118–143, 1995b.

F. D. Davis, "Perceived usefulness, perceived ease of use, and user acceptance of information technology," *MIS Quarterly*, vol. 13(3), pp. 319–340, 1989.

J. Downey, and H. V. Kher, "A longitudinal examination of the effects of computer self-efficacy growth on performance during technology training," *Journal of Information Technology Education*, vol. 14. pp. 91–111, 2015.

U. O. Ekong, P. Ifenedo, C. K. Ayo, and A. Ifenedo, "E-commerce adoption in Nigerian businesses: An analysis using the technology-organization-environmental framework," In *Leveraging Developing Economies with the Use of Information Technology: Trends and Tools*, IGI Global, Hershey, PA, pp. 156–178, 2012.

M. H. Fagan, S. Neill, and B. R. Wooldridge, "An empirical investigation into the relationship between computer self-efficacy, anxiety, experience, support and usage," *Journal of Computer Information Systems*, vol. 44(2), pp. 95–105, 2004.

Y. S. Foon, and B. C. Y. Fah, "Internet banking adoption in Kuala Lumpur: An application of UTAUT model," *International Journal of Business and Management*, vol. 6(4), pp. 161–167, 2011.

P. Galan, M. Giraud, and L. Meyer-Waarden, "A theoretical extension of the Technology Acceptance Model (TAM) to explain the adoption and the usage of new digital services," *Annual conference of the European Marketing Academy*, 42, Istanbul, 2013.

D. Gefen, W. Straub, and M. C. Boudreau, "Structural equation modeling and regression: Guidelines for research practice," *Communication of the Association for Information Systems*, vol. 4(1), pp. 1–77, 2000.

J. F. J. Hair, W. C. Black, B. J. Babin, R. E. Anderson, and R. L. Tatham, *Multivariate data analysis*, Prentice Hall, Upper Saddle River, NJ, 2010.

L. T. Hu, and P. M. Bentler, "Cutoff criteria for fit indexes in covariance structure analysis: Conventional criteria versus new alternatives," *Structural Equation Modeling: A Multidisciplinary Journal*, vol. 6(1), pp. 1–55, 1999.

S. H. Im, and M. S. Kang, "An international comparison of technology adoption: Testing the UTAUT model," *Information & Management*, vol. 48(1), pp. 1–8, 2011.

ITU, "The internet of Things, International Telecommunication Union (ITU), Internet Report," available from http://www.itu.int/dms_pub/itu-s/opb/pol/S-POL-IR.IT-2005-SUM-PDF-E.pdf, 2013.

A. Kesharwani, and S. S. Bisht, "The impact of trust and perceived risk on internet banking adoption in India: an extension of technology acceptance model," *International Journal of Bank Marketing*, vol. 30(4), pp. 303–322, 2012.

D. Kirtsis, "Closed-loop PLM for intelligent products in the era of the internet of things," *Computer-Aided Design*, vol. 43(5), pp. 479–501, 2011.

R. B. Kline, *Principles and practice of structural equation modeling*, The Guilford Press, New York, NY, 2010.

R. Kranenburg, and E. Anzelmo, "The internet of things, 1st Berlin symposium on internet and society," Oct25–27, 2011.

M. C. Lee, "Factors influencing the adoption of internet banking: An integration of TAM and TPB with perceived risk and perceived benefit," *Electronic Commerce Research and Applications*, vol. 8(3), pp. 130–141, 2009.

Z. Ma, L. Ma, and J. Zhao, "Evidence on E-banking quality in the China commercial bank sector," *Global Journal of Business Research*, vol. 5(2), pp. 73–83, 2010.

G. M. Marakas, M. Y. Yi, and R. D. Johnson, "The multilevel and multifaceted character of computer self-efficacy: Toward clarification of the construct and an integrative framework for research," *Information Systems Research*, vol. 9(2), pp. 126–163, 1998.

C. Martins, T. Oliveira, and A. Popovič, "Understanding the Internet banking adoption: A unified theory of acceptance and use of technology and perceived risk application," *International Journal of Information Management*, vol. 34(1), pp. 1–13, 2014.

G. C. Moore, and I. Benbasat, "Development of an instrument to measure the perceptions of adopting an information technology innovation," *Information Systems Research*, vol. 2(3), pp. 173–191, 1991.

A. Murphy, D. Coover, and S. V. Owen, "Development and validation of the computer self-efficacy scale," *Educational and Psychological Measurement*, vol. 49(4), pp. 893–899, 1989.

J. Pallant, *SPSS survival manual: A step-by-step guide to data analysis using SPSS* (4th ed.), Open University Press, Berkshire, 2010.

K. Pretz, "The Next Evolution of the Internet," available from http://theinstitute.ieee. org/technology-focus/technology-topic/the-next-evolution-of-the-internet, 2013.

E. M. Rogers, *Diffusion of innovations* (4th ed.), Free Press, New York, 1995.

R. E. Schumacker, and R. G. Lomax, *A beginner's guide to structural equation modeling, Routledge*, Lawrence Erlbaum, York, 2010.

G. Tabachnick, and L. S. Fidell, *Using multivariate statistics* (5th ed.), Allyn & Bacon, Needham Height, MA, 2007.

A. Tarhini, K. Hone, and X. Liu, "Factors affecting students' acceptance of e-learning environments in developing countries: A structural equation modeling approach," *International Journal of Information and Education Technology*, vol. 3(1), pp. 54–59, 2013a.

A. Tarhini, K. Hone, and X. Liu, "User acceptance towards web-based learning systems: Investigating the role of social, organizational and individual factors in European higher education," *Procedia Computer Science*, vol. 17(1), pp. 189–197, 2013b.

A. Tarhini, T. Teo, and T. Tarhini, "A cross-cultural validity of the E-learning acceptance measure (ElAM) in Lebanon and England: Aconfirmatory factor analysis," *Education and Information Technologies*, vol. 21(5), pp. 1269–1282, 2015.

K. H. Tarhini, and X. Liu, "The effects of individual differences on e-learning users' behavior in developing countries: Astructural equation model," *Computers in Human Behavior*, vol. 41(1), pp. 153–163, 2014.

G. Tornatzky, and K. J. Klein, "Innovation characteristics and innovation adoption implementation: A meta-analysis of findings. *IEEE Transactions on Engineering Management*," vol. 29(1), pp. 28–45, 2012.

M. Tu, "An exploratory study of internet of things (IoT) adoption intention in logistics and supply chain management," *The International Journal of Logistics Management*, vol. 29(1), pp. 131–151, 2018.

F. C. Tung, and F. C. Chang, "Nursing student's behavioral intention to use online courses: Aquestionnaire-based survey," *International Journal of Nursing Studies*, vol. 45(9), pp. 129–309, 2008.

V. Venkatesh, M. G. Morris, G. B. Davis, and F. D. Davis, "User acceptance of information technology: Toward a unified view," *MIS Quarterly*, vol. 27(3), pp. 425–478, 2003.

V. Venkatesh, J. Thong, and X. Xu, "Consumer acceptance and use of information technology: Extending the unified theory of acceptance and use of technology," *MIS Quarterly*, vol. 36(1), pp. 157–178, 2012.

V. Venkatesh, and X. Zhang, "Unified theory of acceptance and use of technology: US vs. China," *Journal of Global Information Technology Management*, vol. 13(1), pp. 5–27, 2010.

R. H. Weber, "Internet of things–new security and privacy challenges," *Computer Law & Security Review*, vol. 26(1), pp. 23–30, 2010.

F. Welch, "Education in production," *Journal of Political Economy*, vol. 78(1), pp. 35–59, 1970.

M. Y. Yi, J. D. Jackson, J. S. Park, and J. C. Probst, "Understanding information technology acceptance by individual professionals: Toward an integrative view," *Information & Management*, vol. 43(3), pp. 350–363, 2006.

H. S. Yoon, and L. M. B. Steege, "Development of a quantitative model of the impact of customers' personality and perceptions on internet banking use," *Computers in Human Behavior*, vol. 29(3), pp. 1133–1141, 2013.

S. Y. Yousafzai, "A literature review of theoretical models of internet banking adoption at the individual level," *Journal of Financial Services Marketing*, vol. 17(3), pp. 215–226, 2012.

C. S. Yu, "Factors affecting individuals to adopt mobile banking: Empirical evidence from the UTAUT model," *Journal of Electronic Commerce Research*, vol. 13(2), pp. 104–121, 2012.

Y. Y. Yuen, P. Yeow, N. Lim, and N. Saylani, "Internet banking adoption: Comparing developed and developing countries," *Journal of Computer Information Systems*, vol. 51(1), pp. 52–61, 2010.

T. Zhou, Y. Lu, and B. Wang, "Integrating TTF and UTAUT to explain mobile banking user adoption," *Computers in Human Behavior*, vol. 26(4), pp. 760–767, 2010.

Chapter 7

Intelligent approaches for disease detection and prevention

Saumya Yadav
Indraprastha Institute of Information Technology (IIIT), Delhi, India

Deepak Chandra Joshi
Shiv Nadar University, Ghaziabad, India

Abhishek Joshi
Sungkyunkwan University, Seoul, South Korea

Sanjay Mathur and Manoj Bhatt
College of Technology, G.B.P.U.A&T., Pantnagar, India

CONTENTS

7.1 INTRODUCTION

The healthcare sector is presently experiencing the implementation of abundant innovations. Human health being the most important aspect to consider, researchers have put a lot of efforts in developing various new medicines, devices and systems to detect, prevent and treat different diseases. Currently, there are two main concerns in the medical sector: increased health maintenance expenditures and a shortage of medical experts. According to the World Health Organizations (WHO), in 2013 the worldwide demand and actual number of health staff were 60.4 million and 43 million, respectively [1]. By 2030, it is estimated that these numbers will rise to a demand for 81.8 million with an actual number of staff of 67.3 million. The shortage of medical amenities and medical experts is still a serious problem. Worldwide, by 2050 there will be a considerable increase in the population of adults above 60 years of age [2] due to the continuing decrease in the birth rate, a trend which will persist in the coming years. With the increase of age, an individual is more likely to get infected with different viruses and diseases which require long-term care for them to return to health. Ultimately, the increased human resources, medicines, medical devices, and expenditure will be required by the population. The constant evolution of new viruses and diseases makes it challenging to stay ahead of the curve, but with the help of different approaches to the treatment of information, it is possible to conquer these challenges.

Across the globe countries devote a considerable proportion of their gross national product (GNP) is spent on public healthcare. A WHO report exhibited that the related figures in India, Australia, Russia, Canada, the U.S.A., and China were 4.7%, 9.4%, 7.1%, 10.5%, and 5.6% respectively [3] and it is likely that these figures will increase in the coming years. In order to optimize efficiency, the world's healthcare systems requires an intelligent solution which gives accurate results in real time. These intelligent solutions can be achieved with the help of AI, IoT, and blockchain. Integrating all these technologies with medical data will undoubtedly help in resolving healthcare problems. Taken together, these systems can help medical specialists to identify the disease by collecting information from the patient using different sensors, comparing the symptoms of the disease with previous patients using blockchain and detect the disease through the use of AI techniques. Such techniques are supportive to medical experts in detecting the disease at the early stages of development, reaching the outcome of complex disease in an easy and faster way. In addition, these techniques support the patients at reduced charge, take lesser time and make an accurate analysis about the probable disease. As a large amount of healthcare data is available today and new advanced data analytics techniques made it possible to develop accurate medical diagnostic tools, primarily to help a lot of people in remote locations which lack medical professionals. AI is merged with the IoT to monitor the health conditions of the patient using different sensors and giving the output in real time. It aims to optimize the time and financial status of a person without compromising the health of a person. Additionally, it provides support to medical staff and reduces government spending in the health sector.

Bearing in mind the importance of this important evolution in healthcare systems, this work presents a detailed study of different technologies

employed in the healthcare sector. The chapter includes various works performed by the researchers in the medical field using machine learning, IoT,
and blockchain technology. The chapter further aims to look at different
researches performed for detecting, preventing, and treating diseases such as
diabetes mellitus (DM), cardiovascular disease, chronic kidney disease
(CKD), and coronavirus using machine learning, the IoT, and blockchain
technology. The structure of the chapter is as follows: section 7.2 discusses
AI, the IoT, and blockchain technology; and section 7.3 presents the research
work done in the healthcare sector to prevent, detect and treat different
diseases using these technologies. Finally, section 7.4 concludes the chapter
and discuss the future possibility of these technologies in healthcare.

7.2 ARTIFICIAL INTELLIGENCE, INTERNET OF THINGS, AND BLOCKCHAIN TECHNOLOGY

7.2.1 Artificial intelligence

AI is a subsection of computer science that tries to develop intelligent
machine and technologies. It makes efforts to program machines and implement human intelligence in them to impersonate their actions. The prob-
lem-solving and learning approach is being evolved in machines to make
them more comprehensive. John McCarthy coined the term "artificial intelligence" in 1956 during his workshop. The goal of this was "to proceed on
the basis of the conjecture that every aspect of learning or any other feature
of intelligence can in principle be so precisely described that a machine can
be made to simulate it". Machine learning (ML) is a subsection of AI developed by Arthur Samuel in 1959 to develop various algorithms for extraction
of the information from the input data and the exploration of generalised
characteristics in the data points. These characteristics are general mathematical models established with certain rules and principles. The models
and decision process is generated from the input data and human guidance.
Once developed, ML automates the process using mathematical models and
programming that eliminate the human expert in the process.

Among the existing technologies, AI is the most promising and powerful
technology in the medical field [4–6]. AI can be described as a simulation of
human intelligence in a machine. AI is implemented when the system starts
making decisions like a human being. It is a large field that includes different
techniques such as computer vision, machine learning, and deep learning,
which are widely used in the medical field for performing segmentation,
classification, detection, prevention and treatment [7]. It aims to develop an
automatic system that learns from the input data and solves complex problems. AI uses the mathematical tool, i.e., ML, to iteratively learn from the
input data and tries to find the pattern from these data. On getting the
pattern, it generates a ML model which is utilized by the user to perform

different tasks. AI is hungry for data. The bigger data used for training the model, the more accurate the result it will provide. In order to make use of AI in the healthcare sector, medical data is collected from clinical activities, such as diagnosis, screening, scanning, treatment etc. Various modalities, such as magnetic resonance imaging (MRI), positron emission tomography (PET), ultrasound, microscopy, X-Ray, and computed tomography (CT), are widely used to obtain a fast and accurate medical diagnosis. In addition, they help healthcare experts to make a decision for surgeries and navigating tools while performing surgeries [8].

7.2.2 Machine learning

Researchers are working in the field of AI with systems with abilities such as perception, reasoning and learning. This field is evolved further and applied in different domains and industries for a range of applications, e.g. computer science, self-driving cars, drug development, mathematics, finance, production, automobile, and many more. ML algorithms can learn from the data of a similar group of subjects, a connection between subject features and from the outcome. Computational systems are trained through different algorithms and statistical models for the analysis of the sample data, so as to develop an efficient learning process. Domain knowledge is required for the selection of proper features in the ML algorithm for improving the predictions. Its features include the variable or descriptive attributes that are recorded and quantified from raw data to train the ML model. In a broad manner, ML is developing an algorithm or computer program which progressively improve its performance on its own. ML is principally divided into four types: Supervised Learning, Unsupervised Learning, Semi-supervised Learning, and Reinforcement Learning (Figure 7.1).

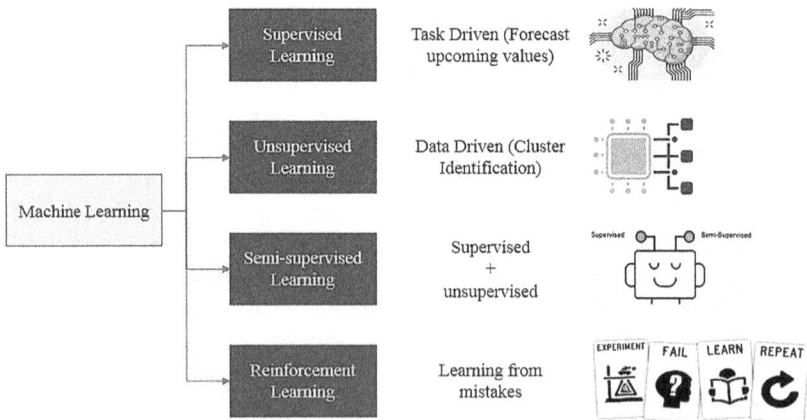

Figure 7.1 Types of ML.

Supervised Learning occurs when the algorithm is trained using the labelled data set. These data sets have both input and output parameters. For supervised learning, both training and validation data sets are labelled. In supervised learning algorithms, the input is a set of N samples, represented as X in Equation 7.1,

$$X = \left[x_i, x_{i+1}, \ldots, x_{N-1} \right] \qquad (7.1)$$

and the value of *i* ranges from 0 to N-1 for all used samples. The term x_i represents the feature vector for the ith sample in the input data. The output class or label is represented by Y corresponding to each input data given in Equation 7.2,

$$Y = \left[y_i, y_{i+1}, \ldots, y_{N-1} \right] \qquad (7.2)$$

The whole training set is represented by *T*, which is the set of input data x_i and output class or label y_i. T is represented as in Equation 7.3,

$$T = \left[(x_i, y_i), (x_{i+1}, y_{i+1}), \ldots, (x_{N-1}, y_{N-1}) \right] \qquad (7.3)$$

where $i = \left[0, 1, 2, \ldots, (N-1) \right]$.

The learning algorithm tries to build a relationship or function *f* which maps *X* to *Y*, as shown in Equation 7.4.

$$f : X \rightarrow Y \qquad (7.4)$$

Steps for solving a problem using a supervised learning are given as in Figure 7.2.

Data analysis: There should be an analysis of the data before training to know what kind of data is to be given, e.g. there may be a different possibility of the training set for optical character recognition. It can be a word or perhaps a single character.

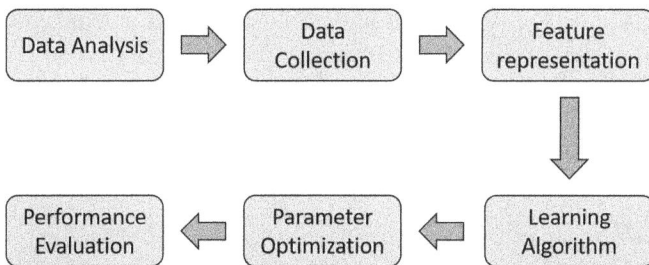

Figure 7.2 Steps for supervised learning.

Data collection: The data is to be collected according to the requirement of the targeted task which should have good representation to the real-world use. The input data is collected with their corresponding labels. The labels can be generated with the expert's opinion or through the conventional method's measurement results.

Feature representation: The training model's performance is significantly dependent on the selection of the features from the input data. The number of the features represents the characteristics of the input object. The number of the features should be optimally selected; it should not be too large or not very low. It should contain several features that could be required for an accurate prediction of the output.

Learning algorithm: The next step is to choose or design the learning methodology for training with input data. The learning should be done depending on the mathematical computation which is related to the performance of the system. An ideal system should be optimised such that it is accurate enough and in the same way computationally inexpensive.

Parameter optimization: Once the algorithm is selected to train with a training set of the input data, the parameters used in the algorithm should be optimized subsequently to enhance the performance of the system. The optimization is done concerning the test set's obtained performance or validation set, which is a kind of unseen data to the training algorithm. Cross-validation is also used to exhaustively evaluate the performance of the system.

Performance evaluation: The final step is to evaluate the performance of the system in a certain set of parameters. The chosen parameter can be different according to the selected problem. The most common parameters are accuracy, specificity, sensitivity, F1-score, kappa, area under the curve, confusion matrices etc.

Various machine-learning algorithms, such as Support Vector Machines (SVM), Random Forest, Decision tree, Naïve Bayes and k-Nearest Neighbour classifiers, are employed to perform different tasks such as population growth prediction, weather forecasting, speech recognition, digit recognition etc. It has been attracted in research for solving complex problems in medical [9–14].

Unsupervised Learning allows the ML model to learn without a labelled data set and with no learning guidance provided. In unsupervised learning, the role of the algorithm is to cluster unprepared and unsorted data according to some similarities and variations with minimum human supervision. It tries to find the variation in the input data and classifies the output based on some differences. The most common unsupervised learning method is clustering and principal component analysis. Steps for unsupervised learning are shown in Figure 7.3.

Principal component analysis and clustering are the two most commonly used probabilistic methods in the domain of unsupervised learning. Clustering

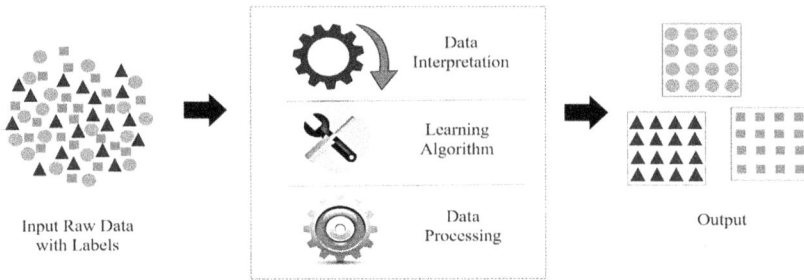

Figure 7.3 Steps for unsupervised learning.

is used to group the unlabelled, unclassified, uncategorised data. It discovers patterns of interest in the natural feature space of the data, such as behavioural analysis of a group of customers. A cluster in a feature space represents the similar data points of the domain which are closer to other groups of the data points or clusters. Generally, clusters have their centroid point or sample in that feature space with similar traits. The cluster may have a centre (the centroid) that is a sample or a point feature space, and may also have a boundary or extent. Clustering is broadly classified into two groups, namely soft clustering and hard clustering. In the case of soft clustering, the data points are assigned with a probability of belongingness to a certain cluster. But the hard clustering gives the binary results, i.e. the given data point may or may not be associated with the cluster. The data point can belong to only one cluster out of all generated clusters in the case of soft clustering. The other two major clustering techniques with wide applications are K-means clustering and hierarchical clustering. The K-means clustering is an iterative method of clustering that works on the principle to find local maxima in each iteration. The term 'K' in K-means clustering represents a number of the clusters generated with this method. Hierarchical clustering works on building the hierarchy of the clusters in which different clusters are generated from the given data points. Then, the two nearest clusters are merged into the same cluster and the algorithms end in a single cluster. It is implemented using the bottom-up approach. The merging of the clusters is done based on different mathematical criteria such as Mahalanobis distance, Euclidean Distance, Manhattan distance, squared Euclidean distance etc. K-means algorithms can handle a large amount of data as compared to hierarchical clustering. The time complexity of hierarchical clustering is quadratic, whereas the time complexity of the K-means algorithm is linear. K-means algorithm works well with the data in situations where the clusters are spherical or circular in shape. Other examples of the clustering methods include Density-Based Spatial Clustering of Applications with Noise (DBSCAN), mean-shift algorithm, agglomerative hierarchical algorithm, affinity propagation, spectral clustering, etc.

The principal component analysis is basically a dimensionality reduction method and is employed in dealing with large-sized data sets. It transforms large size data sets into smaller sets such that most of the information is still preserved. With the aid of the principal component analysis, the number of variables is reduced so that their analysis becomes easier. Although it comes at the cost of compromised accuracy it can make the methodology much simpler and easier to implement. The first step is the standardization of the data so that each variable has an equal contribution to the analysis. This step helps to avoid those situations where variables with larger ranges dominate over data with small range variables. The next step is the computation of the covariance matrix to analyse the variation of the input data from the mean value. In some situations where two or more variables are highly related to each other, then those features are redundant to process and used for dimensionality reduction. Then, in the next step, Eigenvalues and Eigenvectors of the covariance matrix are to be computed and those features will be chosen which are depicting high variance among them. Unsupervised algorithms are widely used to perform several disease detection and localization tasks [15–19].

Semi-supervised learning is a fusion of supervised and unsupervised learning. The training data for semi-supervised learning employs both labelled and unlabelled data sets. In semi-supervised learning, models are first trained with the labelled data set to set some learning rules; then the rules are modified to train with the unlabelled data set. The robustness of such ML models is dependent on the consistency of the labelled data set. The better the labelling of the data set, the more robust the model. One method of semi-supervised learning is to combine the classification and the clustering method to generate a model. Using the clustering method to assemble the input data allows us to secure the most relevant samples and label of the data. Clustering is unsupervised learning, so it does not require any instruction for grouping the data. Later, you can use the classification method on the selected relevant data to generate the prediction model. The healthcare sector uses semi-supervised learning to assist the experts using different technologies [20–26].

Reinforcement Learning is a type of ML whose agents learn from the consequences of its action rather than learning from external resources. It selects its action from past experiences and also by exploring new actions. In this method, an agent tries to achieve a goal in a potentially complex and uncertain environment. Here, the computer tries to solve problems in a manner similar to that of a game. The machine works on the trial-and-error phenomenon to solve a problem and generate the output. The AI-based method gets rewards or penalties according to the performed action. The main target to enhance the amount of the reward which depicts its good chances to take the correct decision. In this type of learning, the model learns on its own and only rewards and penalties are provided. The machine is performing multiple trials to maximize its performance. The main criticality in this kind of system is to prepare the simulation environment in which these tasks will be performed. The system which works in real-world situations become more challenging, e.g. automatic self-driving vehicles, other simulators etc. As there is no communication to the network apart from the reward and penalties system, scaling and controlling the agent is another criticality. In such a situation, the computer goes into a situation called catastrophic forgetting in which the computer's previous knowledge is erased from the network while it trains for new knowledge [27].

Positive and negative reinforcement are two techniques based on the type of reinforcement. If some event occurs due to some behaviour, that will increase the frequency and strength of the behaviour. Thus, this gives a positive trigger to the system, meaning that the system can be sustained longer. Yet these systems can be overloaded with the states due to greater reinforcement which can decrease the performance of the system. It has multiple options to explore before opting for the correct option. It is different from supervised learning as it does not have a labelled training data set to perform correct training [28–32]. Steps for reinforcement learning are shown in Figure 7.4.

In the recent years, a subfield of ML, "deep learning", has provided substantial growth in accuracy, robustness and provides real-time output using new learning algorithms. It comprises a deep graph with different layers performing different tasks. Deep learning layers consist of convolution, pooling, activation, and fully connected layers. These deep layers easily find the complex features from the input medical data to generate a robust model for performing a different task. Various deep models are proposed by the researchers to perform different tasks on the input medical data [33–36]. Computer vision (CV) is another field of AI. It supports computers in identifying and interpreting the visual world. It assists the medical experts by becoming an eye of the medical devices to identify, classify, and detect different diseases from the input modalities of a patient [37]. CV automatically extracts information from the image and with the help of a ML algorithm, it classifies the image content. CV in the healthcare sector enables the experts

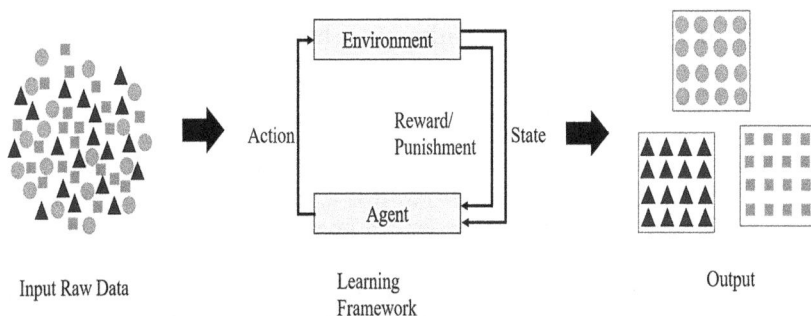

Figure 7.4 Steps for reinforcement learning.

to diagnose their patient, perform surgery, monitor the evolution of different diseases, and detect disease in clinical images, the early identification of diseases and the detected part of the body.

7.2.3 Internet of Things

IoT can be characterized as an interconnection of physical objects or "things", consisting of sensors, processing network and software to connect and exchange the data with other devices and systems over the Internet. Sensors and other connected devices collect the data from the given environment and information are extracted from the raw data. Then the information is transferred to cloud servers and other devices via the Internet. Its applications are able to function with the different domains and enable high analyses and management of complex interactions in the devices [38]. IoT devices, such as wearable sensors, monitors, and smart pills, help to collect the data and ML models can use this data for disease prediction. Different sensors are embedded into the medical system via automated signal processing, automation, and computer technologies. These sensors collect information that allows a clinician to recognize the situation of the patient in real time and to help them by providing treatment. Different medical signals can be found in the form of blood pressure, body temperature, heartbeat rate, electromyograms (EMGs), electrocardiograms (ECGs), electroglottographs (EGGs), electroencephalograms (EEGs), and electrooculograms (EOGs). An IoT-based healthcare monitoring system can use these signals to help the clinician to monitor the patient proactively. A smart gateway assembles data from different smart devices. It can perform processing, compress data, use noise filtering on the medical devices, and analyse the data to detect and predict the risky pattern in a patient's heath. The Internet of Medical Things (IoMT) is a further tweaking of the system. It is a combination of medical applications and systems that connect healthcare data using network technologies. IoMT plays a vital role in developing a smart standalone healthcare system. IoT with the help of AI supports the medical experts to take the medical data of the patient remotely, leading to effective diagnosis. It transfers the data from one device to another to enhance the performance of the system in an automatic manner without any human intervention [39–48].

7.2.4 Blockchain technology

Blockchain technology was introduced in 2008 with the introduction of the Bitcoin cryptocurrency technology. It is a public ledger, tracking assets, distributed, recording transaction, and assure immutability in between a p2p network of computers. Blockchain technology is widely used in various fields, including healthcare, finance, industries, and business. It is an innovative data structure that contains the list of records called a block [46]. Every data block consists of time-stamped batches of the latest transaction, a hash (unique identifier or digital fingerprint), and a hash of the previous data block. Modifying any block of blockchain technology is therefore difficult as changing one block leads to changes in all the blocks behind the modified block. Several features of this technology, such as the immutability of stored data, decentralization, and traceability are attracting the healthcare sector for merging blockchain with medical devices [49–54]. It is expected to upgrade the information management of patients, improve clinical research, insurance claim process, and the management of data records [55]. Blockchain technology enables institute-driven interoperability to transform to patient-centred interoperability. Researchers can access some part of the patient data for a limited time period after taking consent from the patients using blockchain technology. It also allows patients to connect to other hospitals and automatically collect their medical data from the previous hospitals. In addition, blockchain in healthcare is capable of protecting the patient's information [15]. It is constantly upgrading, and it has several challenges which must be resolved to merge it with healthcare and medical applications. The foremost challenge concerns confidentiality and the second challenge scalability and speed.

AI with IoT and blockchain technology is a boon for the healthcare sector. These technologies are helping to boost up the clinical data to create efficient and accurate healthcare treatments for patients and increase productivity in the workforce. These computational tools also help in optimizing the parameters by which the root cause of a disease can be identified. Using IoT sensors, clinicians are collecting, analysing, and reporting the data promptly and sending it to doctors for deeper insights into the health of the patients. Doctors are performing real-time monitoring of patients who are seriously ill. Blockchain is bringing accessibility and transparency in critical patient's information from all the relevant doctors. Also, it is easier for the patient to switch to another hospital using a blockchain network that includes the patient data. Using AI algorithms, medical experts can easily detect, prevent and diagnose the disease in real time. The perfect integration of these three technologies and digital innovations is leading to improvements in the healthcare field. This enhancement brings new options of treatment by medical experts. Technologies also assist the healthcare members to track the patients in the hospital with the help of CV [37] and maintain their routine. Figure 7.5 represents the integration of AI, IoT and blockchain with healthcare.

7.3 DETECTION, PREVENTION AND TREATMENT OF DISEASE

With the advancement in medical treatment, the patient population is increasing with more complex and multiple diseases being detected. This is making medical decisions harder as medication may help one patient and

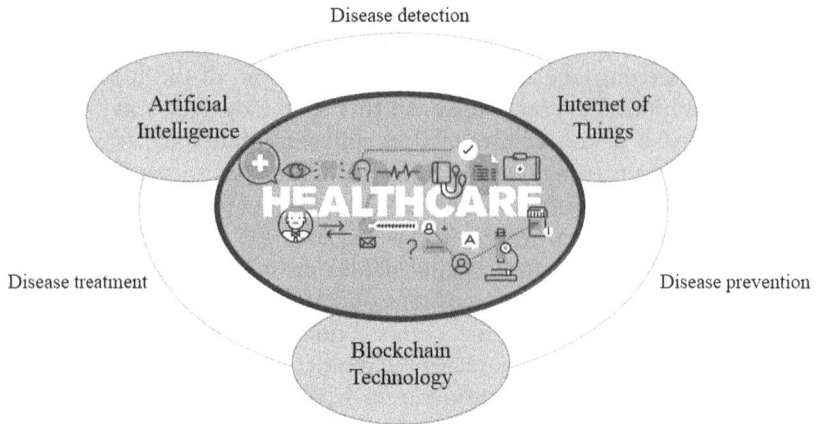

Figure 7.5 AI, IoT and blockchain technology merging with healthcare.

harm another. The latest emerging technologies in healthcare sectors assist medical experts. The researcher used AI algorithms with IoT-based sensors and blockchain technology to perform different tasks such as the detection, prevention and treatment of different diseases.

7.3.1 Diabetes mellitus

Diabetes mellitus (DM) is a complex metabolic disorder that can lead to the damage of multiple body organs. It is also one of the world's primary metabolic diseases [56]. There is an emergent need to develop models/tools for the risk assessment of disease, along with detecting susceptibility and prognosis. The disease is linked to abnormal blood glucose levels in the human body. Type 1 diabetes is the result of inadequate insulin production and Type 2 diabetes is an output of oxidative stress, which results from defective redox reactions and increased reactive metabolites (RMs) [57]. Looking at the high prevalence of DM in the population, it would seem that most of the individuals are undiagnosed. Hence, there is an emergent need to develop a tool to assess the risk assessment of disease, susceptibility and prognosis. In addition, there is a requirement to prevent diabetes by periodically taking measurement and keep track of health data. Various risk assessment and information security models are developed using AI, IoT and blockchain to detect, prevent and cure DM in the patients as shown in Table 7.1.

7.3.2 Cardiovascular Disease (CVD)

According to a WHO report, an annual number of 12 million people die due to heart disease at the global level. Heart disease is the reason for the death of around 31% of the world's population. WHO assumes that by 2030, the

Table 7.1 Comparative analysis of AI+IoT+blockchain technology for DM

Reference	Objective	Method	Accuracy (%)
[58]	Security and privacy	Blockchain and IoT	NA
[59]	Preventing by predicting the disease will occur within 5 years	ML algorithm	83.30
[60]	Prediction of diabetes	Hadoop cluster	94
[61]	Glucose monitoring system for diabetes prevention	IoT	NA
[62]	Detecting the disease	AI algorithm	94.1417
[63]	Prediction of DM	ML algorithm	81

Table 7.2 Comparative analysis of AI+IoT+blockchain technology for CVD

Reference	Objective	Method	Accuracy (%)
[66]	Prediction of heart disease	Autoencoder-based artificial neural network	90
[67]	Heart disease prediction	Supervised learning algorithms	83
[68]	Cuff-Less Blood Pressure Measurement	IoT	NA
[69]	Heart disease prediction	Cluster-based DT learning	85.90
[70]	Heart Rate Monitoring System	IoT	NA
[71]	Heart disease prediction	Hybrid ML Techniques	88.7

death rate due to heart disease will increase up to 23.6 million [64]. The increased death rate and numerous reasons for heart disease make prediction complicated using a conventional method. Researchers are therefore focused on developing new methods for diagnosing heart disease detection, prediction and treatment with the latest technologies. AI predicts heart disease using the attachment of different IoT sensors. It evaluates patient data and predicts the chances of heart disease [65]. Input data required for the heart disease prediction using AI involves time series, images, tabular data, and text data. The input data to be fed into the AI model are the risk factor features for predicting the disease. Data include the weight, heart rate, blood pressure, and physical activities of the respective patients. Some heart disease prediction using AI, and IoT are given in Table 7.2.

7.3.3 Chronic Kidney Disease (CKD)

CKD is a category of kidney disease in which there is a steady loss of glomerular filtration rate (GFR) over a period of more than three months and the intensifying problem of health [72]. It is a highly complex, secret, and

Table 7.3 Comparative analysis of AI+IoT+blockchain technology for kidney disease

Reference	Objective	Method	Accuracy (%)
[74]	Detection and diagnosis of CKD	IoMT with ML	99.7
[75]	Disease classification	ML using multi-kernel	98.5
[76]	Diagnosis of CKD	ML	99.83
[77]	Rule Induction and Prediction of CKD	Hybrid ML Techniques	99.75
[78]	Cuff-Less Blood Pressure Measurement	ML, Internet of Things and Cloud Computing	97.8
[79]	Predicting the Risk of CKD	ML algorithms	NA

progressive condition that focuses on the physical functioning of some organs. According to WHO estimates, the estimated past costs of dialysis treatment stand at $1.1 trillion [73]. It is difficult to predict the timing of failure of the kidney because of its complicated nature, the heterogeneity in different patient's condition, and the absence of early symptoms. If it is diagnosed at an earlier stage, treatment can be started earlier and may reduce the chances of death. Using ML for disease detection can be more accurate, robust, and make a prediction in real time when the model is trained with a clean and suitable data set. The data set for generating a detection or prediction model can be of different modalities. Researchers performed image processing to segment out the kidney image, illustrating all defects. Table 7.3 includes CKD detection and prevention using AI and IoT.

7.3.4 Coronavirus disease (COVID-19)

COVID-19 originated in China. It is a disease which is caused by severe acute respiratory syndrome coronavirus-2 (SARS-CoV-2). It has been found to be a disease that can be severe in patients suffering from other conditions and has a current fatality rate of 2% [80]. This life-threatening COVID-19 pandemic is having a very strong impact on the health of the global population and the number of patients is increasing for individual testing with regard to the disease. Conventionally, medical experts have used a chest x-ray image or CT scan of an individual in order to identify the disease. This leads to a heavy workload and pressure on radiologists to perform the test and give accurate results in real time. The fusion of the conventional method with technologies will reduce the workload and also for the simultaneous performance of multiple tests. An infected patient's modalities may include numerous opacities when compared with healthy modalities. Researchers took advantage of this feature and with the help of AI proposed various models which can detect COVID-19 from the input data. With the help of IoT, a clinician can ensure that the virus-infected patient remains

Table 7.4 Comparative analysis of AI+IoT+blockchain technology for COVID-19

Reference	Objective	Method	Accuracy (%)
[81]	Detection and diagnosis of CKD	DL framework	99.81
[82]	Monitoring the patient	IoT applications	NA
[83]	Social distance measure	IoT applications	NA
[84]	Uniquely tracking donation	Blockchain-based network	NA
[30]	Prediction of COVID-19	Supervised machine learning	94.99
[85]	Diagnostic framework for COVID-19	Deep Learning algorithms	99.81

quarantined for the appropriate time period. The physician can measure body temperature, oxygen rate, and blood pressure using different IoT sensors. In addition, it helps to monitor the patients' health. Successfully implementing these technologies can improve the productivity of healthcare staff, reduce the number of mistakes, cut down on the costs of scans, and reduce the workload. Several such methods are included in Table 7.4.

7.4 CONCLUSION

People around the globe are affected by different types of deadly diseases. Among the most common diseases are DM, cardiovascular disease, CKD, and, most recently, coronavirus disease, all of which are having an adverse effect on the health and economy of the world. Disease detection is the initial step taken by the medical experts on finding any symptoms. With the increasing population, the demand for medical experts is increasing, and so ultimately is brings the workload and pressure. Merging AI, IoT, and blockchain with the medical field reduces the experts' workload and give precise output in real time. AI and ML techniques are used extensively in clinical decision-making. With the help of AI, researchers generated different models for detecting, segmenting, and predicting diseases. Different sensors of IoT are used to collect data from the patients and also to remotely monitor the health of the patients. Blockchain technology is used to secure the patients' information, to permit the researcher to access the limited data, in order to consulting multiple doctors. The proposed work represents different methods employed in the health sector using AI, IoT, and blockchain technologies. It also outlines the methods used for performing different types of work.

In future, the researchers can merge more relevant techniques in healthcare, which will improve disease detection, prevention, and treatment. It will improve performance, ensure that medical services are financially feasible, safeguard future health, and reduce the workload of medical experts.

REFERENCES

[1] World Health Organization, Health workforce requirements for universal health coverage and the sustainable development goals. (human resources for health observer, 17), 2016.

[2] J. R. Beard, et al., "The World report on ageing and health: a policy framework for healthy ageing." *The Lancet*, vol. 387, pp. 2145–2154, 2016.

[3] World Health Organization, "Total expenditure on health as a percentage of gross domestic product (US $). Global Health Observatory", 2017.

[4] S. E. Dilsizian, E. L. Siegel, "Artificial intelligence in medicine and cardiac imaging: harnessing big data and advanced computing to provide personalized medical diagnosis and treatment." *Current Cardiology Reports*, vol. 16(1), pp. 1–8, 2014.

[5] V. L. Patel, N. A. Yoskowitz, J. F. Arocha, "Towards effective evaluation and reform in medical education: a cognitive and learning sciences perspective." *Advances in Health Sciences Education*, vol. 14(5), pp. 791–812, 2009.

[6] S. Jha, E. J. Topol, "Adapting to artificial intelligence: Radiologists and pathologists as information specialists." *JAMA*, vol. 316(22), pp. 2353–2354, 2016.

[7] G. Litjens, et al., "A survey on deep learning in medical image analysis." *Medical Image Analysis* vol. 42, pp. 60–88, 2017.

[8] Y. Shelke, C. Chakraborty Shelke, "Augmented reality and virtual reality transforming spinal imaging landscape: a feasibility study." *IEEE Computer Graphics and Applications*, vol. 41(3), pp. 124–138, 2020.

[9] M. Zhang, et al., "Prediction of virus-host infectious association by supervised learning methods." *BMC Bioinformatics*, vol. 18(3), pp. 143–154, 2017.

[10] Y. Cai, X. Tan, Y. Cai, X. Tan, X. Tan, "Selective weakly supervised human detection under arbitrary poses." *Pattern Recognition*, vol. 65, pp. 223–237, 2017.

[11] A. Singh, R. Kumar, A. Singh, R. Kumar, "Heart disease prediction using machine learning algorithms." *International conference on electrical and electronics engineering (ICE3)*. IEEE, 2020.

[12] E. Choi, A. Schuetz, W. F. Stewart, et al., "Using recurrent neural network models for early detection of heart failure onset." *Journal of the American Medical Informatics Association*, vol. 24(2), pp.361–370, 2017.

[13] D. Ichikawa, T. Saito, W. Ujita, H. Oyama, "How can machine-learning methods assist in virtual screening for hyperuricemia? A healthcare machine-learning approach." *Journal of Biomedical Informatics*, vol. 64, pp. 20–24, 2016.

[14] S. P. Patro, N. Padhy, D. Chiranjevi, "Ambient assisted living predictive model for cardiovascular disease prediction using supervised learning." *Evolutionary Intelligence* vol. (2), pp. 941–969, 2021.

[15] G. Muhammad, "Automatic speech recognition using interlaced derivative pattern for cloud based healthcare system." *Cluster Computing*, vol. 18(2), pp. 795–802, 2015.

[16] F. A. Khan, et al., "A continuous change detection mechanism to identify anomalies in ECG signals for WBAN-based healthcare environments." *IEEE Access*, vol. 5, pp. 13531–13544, 2017.

[17] S. Lim, C. S. Tucker, S. Kumara, "An unsupervised machine learning model for discovering latent infectious diseases using social media data." *Journal of Biomedical Informatics*, vol. 66, pp. 82–94, 2017.

[18] A. Fong, et al., "Identifying influential individuals on intensive care units: using cluster analysis to explore culture." *Journal of Nursing Management*, vol. 25(5), pp. 384–391, 2017.

[19] K. Bhatia, R. Syal, "Predictive analysis using hybrid clustering in diabetes diagnosis." *Recent Developments in Control, Automation & Power Engineering (RDCAPE)*. IEEE, 2017.

[20] S. R. Jonnalagadda, "A semi-supervised learning approach to enhance health care community–based question answering: A case study in alcoholism." *JMIR Medical Informatics*, vol. 4(3), pp. e5490, 2016.

[21] A. Albalate, W. Minker, *Semi-supervised and unsupervised machine learning: novel strategies*. John Wiley & Sons, 2013.

[22] H. Wu, et al., "Automated comprehensive adolescent idiopathic scoliosis assessment using MVC-Net." *Medical Image Analysis*, vol. 48 pp. 1–11, 2018.

[23] Y. Wang, et al., "Semi-supervised classification learning by discrimination-aware manifold regularization." *Neurocomputing*, vol. 147, pp. 299–306, 2015.

[24] L. Nie, et al., "Bridging the vocabulary gap between health seekers and health-care knowledge." *IEEE Transactions on Knowledge and Data Engineering*, vol. 27(2), pp. 396–409, 2014.

[25] L. Jin, et al., "Integrating human mobility and social media for adolescent psychological stress detection." *International conference on database systems for advanced applications*. Springer, Cham, 2016.

[26] R. A. R. Ashfaq, et al., "Fuzziness based semi-supervised learning approach for intrusion detection system." *Information Sciences*, vol. 378, pp. 484–497, 2017.

[27] C. Kaplanis, M. Shanahan, C. Clopath, "Continual reinforcement learning with complex synapses." *International Conference on Machine Learning*. PMLR, 2018.

[28] M. F. Zohora, et al., "Forecasting the risk of type ii diabetes using reinforcement learning." *2020 Joint 9th International Conference on Informatics, Electronics & Vision (ICIEV) and 2020 4th International Conference on Imaging, Vision & Pattern Recognition (icIVPR)*. IEEE, 2020.

[29] D. Zhang, B. Chen, S. Li, "Sequential conditional reinforcement learning for simultaneous vertebral body detection and segmentation with modeling the spine anatomy." *Medical Image Analysis*, vol. 67, pp. 101861, 2021.

[30] L. J. Muhammad, E. A. Algehyne, S. S. Usman, "Supervised machine learning models for prediction of COVID-19 infection using epidemiology dataset." *SN Computer Science*, vol. 2, p. 11, 2021.

[31] J. Wang, et al., "Deep reinforcement active learning for medical image classification." *International Conference on Medical Image Computing and Computer-Assisted Intervention*. Springer, Cham, 2020.

[32] Z. Liu, et al., "Deep reinforcement learning with its application for lung cancer detection in medical Internet of Things." *Future Generation Computer Systems*, vol. 97, pp. 1–9, 2019.

[33] Y. Cao, et al., "Improving tuberculosis diagnostics using deep learning and mobile health technologies among resource-poor and marginalized communities." *2016 IEEE first international conference on connected health: applications, systems and engineering technologies (CHASE)*. IEEE, 2016.

[34] S. Mishra, A. Dash, L. Jena, "Use of deep learning for disease detection and diagnosis." *Bio-inspired neurocomputing*. Springer, Singapore. pp. 181–201, 2021.

[35] J. Venugopalan, et al., "Multimodal deep learning models for early detection of Alzheimer's disease stage." *Scientific Reports.* vol. 11(1), pp. 1–13, 2021.

[36] J. Wang, et al., "Detecting cardiovascular disease from mammograms with deep learning." *IEEE Transactions on Medical Imaging*, vol. 36(5), pp. 1172–1181, 2017.

[37] R. C. Joshi, et al., "Object detection, classification and tracking methods for video surveillance: a review." *2018 4th International Conference on Computing Communication and Automation (ICCCA)*. IEEE, 2018.

[38] B. D. Martino, et al., "Internet of things reference architectures, security and interoperability: a survey." *Internet of Things*, vol. 1, pp. 99–112, 2018.

[39] H. Sattar, et al., "An IoT-based intelligent wound monitoring system." *IEEE Access*, vol. 7, pp. 144500–144515, 2019.

[40] S. Muthukumar, et al., "Smart Humidity Monitoring System for Infectious Disease Control." *2019 International Conference on Computer Communication and Informatics (ICCCI)*. IEEE, 2019.

[41] F. Miao, et al., "Multi-sensor fusion approach for cuff-less blood pressure measurement." *IEEE Journal of Biomedical and Health Informatics*, vol. 24(1), pp. 79–91, 2019.

[42] F. Yang, et al., "Multi-method fusion of cross-subject emotion recognition based on high-dimensional EEG features." *Frontiers in Computational Neuroscience*, vol. 13, pp. 53, 2019.

[43] Q. Gu, et al., "Health and safety situation awareness model and emergency management based on multi-sensor signal fusion." *IEEE Access*, vol. 7, pp. 958–968, 2018.

[44] M. Muzammal, et al., "A multi-sensor data fusion enabled ensemble approach for medical data from body sensor networks." *Information Fusion*, vol. 53, pp. 155–164, 2020.

[45] K. Lin, et al., "Multi-sensor fusion for body sensor network in medical human–robot interaction scenario." *Information Fusion*, vol. 57, pp. 15–26., 2020.

[46] A. Shahnaz, U. Qamar, A. Khalid, "Using blockchain for electronic health records." *IEEE Access*, vol. 7, pp. 147782–147795, 2019.

[47] J. Chen, et al., "A novel medical image fusion method based on Rolling Guidance Filtering." *Internet of Things*, vol. 14, pp. 100172, 2020.

[48] D. Fabiano, S. Canavan. "Emotion recognition using fused physiological signals." *2019 8th International Conference on Affective Computing and Intelligent Interaction (ACII)*. IEEE, 2019.

[49] S. Tanwar, K. Parekh, R. Evans. "Blockchain-based electronic healthcare record system for healthcare 4.0 applications." *Journal of Information Security and Applications*, vol. 50, pp. 102407, 2020.

[50] K. Shuaib, et al., "Blockchains for secure digitized medicine." *Journal of Personalized Medicine*, vol. 9(3), pp. 35, 2019.

[51] T. Zhou, X. Li, H. Zhao, "Med-PPPHIS: blockchain-based personal healthcare information system for national physique monitoring and scientific exercise guiding." *Journal of Medical Systems*, vol. 43(9), pp. 1–23, 2019.

[52] H. Kaur, et al., "A proposed solution and future direction for blockchain-based heterogeneous medicare data in cloud environment." *Journal of Medical Systems*, vol. 42(8), pp. 1–11, 2018.

[53] E. A. Breeden, C. Davidson, T. K. Mackey. "Leveraging blockchain technology to enhance supply chain management in healthcare: an exploration of

challenges and opportunities in the health supply chain." *Blockchain Healthc Today*, 2018.

[54] V. Patel, "A framework for secure and decentralized sharing of medical imaging data via blockchain consensus." *Health Informatics Journal*, vol. 25(4), pp. 1398–1411, 2019.

[55] T. T. Kuo, H. E. Kim, L. O. Machado, "Blockchain distributed ledger technologies for biomedical and health care applications." *Journal of the American Medical Informatics Association*, vol. 24(6), pp. 1211–1220, 2017.

[56] S. Kang et al., "An efficient and effective ensemble of support vector machines for anti-diabetic drug failure prediction." *Expert Systems with Applications*, vol. 42(9), pp. 4265–4273, 2015.

[57] M. Bhattacharyya, et al., "Oxidative stress-related genes in type 2 diabetes: association analysis and their clinical impact." *Biochem Genet*, vol. 53(4–6), pp. 93–119, 2015.

[58] K. Azbeg, et al., "Blockchain and IoT for security and privacy: A platform for diabetes self-management." *2018 4th international conference on cloud computing technologies and applications (Cloudtech)*. IEEE, 2018.

[59] N. Singh, P. Singh, "Stacking-based multi-objective evolutionary ensemble framework for prediction of diabetes mellitus." *Biocybernetics and Biomedical Engineering*, vol. 40, pp. 1–22, 2020.

[60] N. Yuvaraj, K. R. SriPreethaa, "Diabetes prediction in healthcare systems using machine learning algorithms on Hadoop cluster." *Cluster Computing*, vol. 22(1), pp. 1–9, 2019.

[61] F. Valenzuela, et al., "An IoT-based glucose monitoring algorithm to prevent diabetes complications." *Applied Sciences*, vol. 10(3), pp. 921, 2020.

[62] M. Shuja, S. Mittal, M. Zaman. "Effective prediction of type ii diabetes mellitus using data mining classifiers and SMOTE." *Advances in computing and intelligent systems*. Springer, Singapore, pp. 195–211, 2020.

[63] H. M. Deberneh, I. Kim, "Prediction of Type 2 diabetes based on machine learning algorithm." *International Journal of Environmental Research and Public Health*, vol. 18(6), pp. 3317, 2021.

[64] C. Dangare, S. Apte, "A data mining approach for prediction of heart disease using neural networks." *International Journal of Computer Engineering and Technology (IJCET)*, vol. 3(3), pp. 30–40, 2012.

[65] Y. Yan, et al., "The primary use of artificial intelligence in cardiovascular diseases: what kind of potential role does artificial intelligence play in future medicine?" *Journal of Geriatric Cardiology: JGC*, vol. 16(8), p. 585, 2019.

[66] I. D. Mienye, Y. Sun, Z. Wang, "Improved sparse autoencoder based artificial neural network approach for prediction of heart disease." *Informatics in Medicine Unlocked*, vol. 18, p. 100307, 2020.

[67] M. Chakarverti, S. Yadav, R. Rajan, "Classification technique for heart disease prediction in data mining." *2019 2nd International Conference on Intelligent Computing, Instrumentation and Control Technologies (ICICICT)*.

[68] M. Simjanoska, et al., "Multi-level information fusion for learning a blood pressure predictive model using sensor data." *Information Fusion*, vol. 58, pp. 24–39, 2020.

[69] G. Magesh, P. Swarnalatha. "Optimal feature selection through a cluster-based DT learning (CDTL) in heart disease prediction." *Evolutionary Intelligence*, vol. 14(2), pp. 583–593, 2021.

[70] J. N. Kalshetty, P.M. Varghese, K. Karthik, R. Raj, N. Yadav, "IoT-Based Heart Rate Monitoring System", In *Advances in Artificial Intelligence and Data Engineering: Select Proceedings of AIDE 2019*, pp. 1465–1475, Singapore: Springer Singapore, 2020.

[71] S. Mohan, C. Thirumalai, G. Srivastava. "Effective heart disease prediction using hybrid machine learning techniques." *IEEE Access*, vol. 7, pp. 81542–81554, 2019.

[72] P. E. Stevens, A. Levin, Kidney Disease: Improving Global Outcomes Chronic Kidney Disease Guideline Development Work Group Members*, "Evaluation and management of chronic kidney disease: synopsis of the kidney disease: improving global outcomes 2012 clinical practice guideline." *Annals of Internal Medicine*, vol. 158(11), pp. 825–830, 2013.

[73] M. J. Lysaght, "Maintenance dialysis population dynamics: current trends and long-term implications." *Journal of the American Society of Nephrology*, pp. S37–S40, 2002.

[74] F. Ma, et al., "Detection and diagnosis of chronic kidney disease using deep learning-based heterogeneous modified artificial neural network." *Future Generation Computer Systems*, vol. 111, pp. 17–26, 2020.

[75] H. Fang, et al., "An improved arithmetic optimization algorithm and its application to determine the parameters of support vector machine." *Mathematics*, vol. 10(16), pp. 2875, 2022.

[76] J. Qin, et al., "A machine learning methodology for diagnosing chronic kidney disease." *IEEE Access*, vol. 8, pp. 20991–21002, 2019.

[77] S. H. Ripon, "Rule induction and prediction of chronic kidney disease using boosting classifiers, Ant-Miner and J48 Decision Tree." *2019 international conference on electrical, computer and communication engineering (ECCE)*. IEEE, 2019.

[78] A. Abdelaziz, et al., "A machine learning model for predicting of chronic kidney disease based internet of things and cloud computing in smart cities." *Security in smart cities: models, applications, and challenges*. Springer, Cham, pp. 93–114, 2019.

[79] W. Wang, G. Chakraborty, B. Chakraborty, "Predicting the risk of chronic kidney disease (CKD) using machine learning algorithm." *Applied Sciences*, vol. 11(1), p. 202, 2021.

[80] F. Wu, et al., "A new coronavirus associated with human respiratory disease in China." *Nature*, vol. 579(7798), pp. 265–269, 2020.

[81] A. I. Khan, J. L. Shah, M. M. Bhat, "CoroNet: a deep neural network for detection and diagnosis of COVID-19 from chest x-ray images." *Computer Methods and Programs in Biomedicine*, vol. 196, p. 105581, 2020.

[82] R. P. Singh, et al., "Internet of things (IoT) applications to fight against COVID-19 pandemic." *Diabetes & Metabolic Syndrome: Clinical Research & Reviews*, vol. 14(4), pp. 521–524, 2020.

[83] M. Gupta, M. Abdelsalam, S. Mittal, "Enabling and enforcing social distancing measures using smart city and its infrastructures: a COVID-19 Use case." arXiv preprint arXiv, vol. 2004(09246). 2020.

[84] N. Kamalakshi, "Role of blockchain in tackling and boosting the supply chain management economy post COVID-19." *Convergence of internet of things and blockchain technologies*. Springer, Cham, pp. 193–205, 2022.

[85] S. T. Ahmed, S. M. Kadhem, "Using machine learning via deep learning algorithms to diagnose the lung disease based on chest imaging: a survey." *International Journal of Interactive Mobile Technologies*, vol. 15(16), pp. 95–112, 2021.

Chapter 8

Interoperability in IoT-driven smart buildings

Employing Rule-based decision support systems

Mohan Krishna S.
Alliance University, Bangalore, India

Thinagaran Perumal
University of Putra, Selangor, Malaysia

Sumukh Surya
Bosch Global Software Technologies, Bangalore, India

Chandrashekar
L & T Technological Services, Bangalore, India

CONTENTS

8.1 INTRODUCTION: BACKGROUND AND DRIVING FORCES

Sustainable energy development and climate change are closely linked, and it is increasingly felt across the world that energy consumption must be optimized, and that energy efficiency must be increased. This is applicable to all sectors, but there is greater thrust on the demand side. Commercial buildings

DOI: 10.1201/9781003407300-8

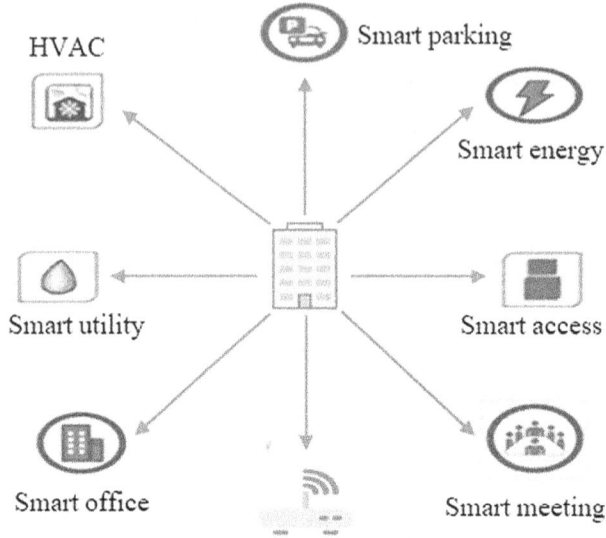

Figure 8.1 Components of smart (intelligent) buildings.

are potentially large energy guzzlers. The power of technology and the internet was harnessed to ensure the building sector optimizes energy use. The rapid spread of the internet and automation across the world, combined with increasingly affordable and high-performance sensors, has led to the concept of intelligent buildings [1–3]. The quest for making commercial buildings and homes intelligent has opened a whole new world of possibilities and challenges with respect to the technology, economy, and environment. The different components of a smart or intelligent building are succinctly represented in Figure 8.1. From the technological point of view, the presence of a multitude of heterogeneous subsystems in a building, along with associated sensor and communication technologies, have also thrown up issues related to coordination and interoperability. From the economic point of view, this requires heavy capital investment, however, the return on investment (ROI) is encouraging with payback periods of less than five years. The environmental degradation is contained as the carbon footprint of the buildings is reduced drastically (with some of them even achieving net zero emissions).

Therefore, it can be safely interpreted that the concept of intelligent buildings is here to stay and, if implemented and executed properly, it will have a cascading effect on the energy patterns of the population and ensure the achievement of the United Nations Sustainable Development Goals (Goal 7) related to energy security and sustainable energy development. This chapter throws light on the technological aspects of intelligent buildings with an emphasis on coordination.

8.2 THE NECESSITY OF DECISION SUPPORT INBUILDING ENERGY SUBSYSTEMS

A commercial intelligent building is made up of several heterogeneous subsystems, namely:

1. Heating, ventilation and air conditioning (HVAC) and energy management subsystems (comprising controllers and sensors)
2. Audio and video subsystems (consoles and speakers)
3. Access control subsystems (door access)
4. Fire management subsystems (alarms)
5. Digital surveillance subsystems (Internet protocol camera, closed circuit television)
6. Water subsystems (valves and pipelines)
7. Other external subsystems and gateways

A designer for intelligent buildings must be able to integrate all these subsystems in a seamless way such that they can easily talk to (coordinate with) each other [4–6].

This is where the issue of interoperability of all these subsystems emerges. The proliferation of internet and automation, low cost of sensors and embedded system technologies and the availability of highly versatile communication infrastructure have revolutionized subsystems inside buildings and made them intelligent. This resulted in increased data intensity of the buildings as huge amounts of data from these subsystems need to be processed, monitored, and analyzed. Figure 8.2 summarizes the different data

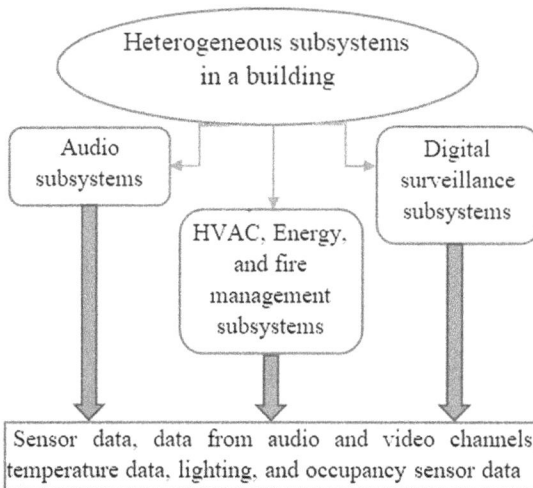

Figure 8.2 Data types for heterogeneous subsystems in a building.

Table 8.1 Pitfalls of Heterogeneous Building Subsystems

Pitfalls	Issues
Integration complexity	There are different resources, platforms and protocols which contributed to the diversity in the BEMS.
Gateways	Although gateways (at the building server infrastructure) were employed for integrating different subsystems, their inherent disadvantage was limited information exchange and not providing full functionality. Besides this, they also created issues related to trouble shooting and support.
Coordination problems	Coordination problems arise and there is a necessity for the various subsystems to perform interoperationally to realize a particular building operation. The data exchange must be seamless and timely.

types employed for some heterogeneous subsystems in a building. The integration complexity also increased and posed several challenges for the management of subsystems in an intelligent building. Table 8.1 provides an overview of the pitfalls involved.

To address these challenges, there is a necessity to have a mechanism which monitors and manages all the subsystems in the building. The BEMS automatically controls the different subsystems by means of real-time data monitoring. The DSS supplements the BEMS and is instrumental in enabling building owners to take decisions related to energy usage and costs. A DSS enables an intelligent building to perceive, learn and adapt. Keeping this in mind, considerable research focused on the development of DSS along with building management.

8.3 DECISION SUPPORT MODEL FOR BEMS

A DSS is a particular class of computerized information systems which is employed for the support of business and decision-making activities of an organization. It is instrumental in the collection, organization and analysis of data and facilitates optimal decision-making and aids the management/facility in operations and planning. DSS is employed in many capital intensive and energy intensive applications like:

Energy management in industries (all sectors)
Financial decision-making in banking and engineering projects
Regulation and Policy making (all sectors)

The basic components of a DSS for BEMS are illustrated in Figure 8.3.

The data can be in the form of raw data, manuals and documents, knowledge inputs from employees and other stakeholders and business models.

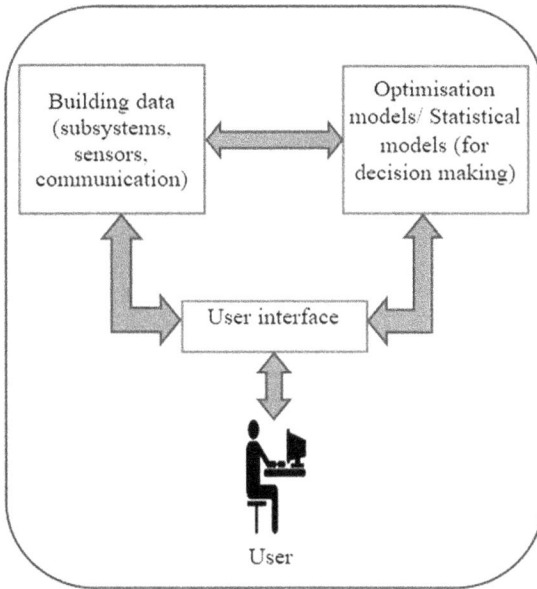

Figure 8.3 Basic components of a DSS for BEMS.

For a building, the data can pertain to different subsystems of the building, the sensor and communication data, building inventory data, data sheets, energy bills, energy consumption data from the star labels of equipment's and buildings and the like. Table 8.2 details the different types of data types employed for heterogeneous subsystems. Before analyzing the impact of a DSS on different building subsystems and the different rule bases which are employed, there are certain ambiguities associated with a DSS used in BEMS which need to be clarified. Table 8.3 summarizes the characteristics of a DSS and addresses certain anomalies pertaining to the use of them.

Table 8.2 Data types employed for heterogeneous subsystems

Building subsystems	Different datatypes employed
Audio and video subsystems	Sensor data, video and audio channels, audio and video flags, control flags for panning, tilting, and zooming etc.
Heating, ventilation, and air-conditioning subsystems	Temperature and thermal management data, energy consumption data etc.
Energy management subsystems	Energy analytics data, energy storage data, energy use data etc.
Surveillance subsystems	Occupational sensor data, time stamp data etc.
Fire management subsystems	Hazard sensor data, equipment data etc.
Lighting subsystems	Occupational sensor data, photo sensor data, luminous efficacy data etc.

Table 8.3 Characteristics of a DSS for BEMS

Characteristics and anomalies associated	
Facilitates decision-making	The DSS must only support and not automate the decision-making process
Interaction and ancillary support	Supports interaction between user and the different layers of IoT-based BEMS. Also supports repeated usage or for ad hoc decision-making
Task-oriented	Have capabilities which provide support for a single or multiple task which may include data intelligence and analysis, identification, design and selection of alternatives and implementation
Identifiable with information systems	It can either be an independent system or a subset of a large integrated information system
Impact on decision-making	Designed to incorporate more accuracy, timeliness, quality, and effectiveness in decision-making

The main purpose therefore is to provide relevant information by the appropriate classification and categorization of data. The management and monitoring of subsystems in a smart building has always been a complex feature due to the heterogeneity and lack of interoperability. A DSS is the need of the hour to manage everyday energy use and ensure optimal quality of life. Rule-based mechanisms are also employed to foresee the logic of the energy management model. Web services and other associated middleware components also significantly integrate the different heterogeneous subsystems (from different manufacturers and operating platforms). There is a distinct difference here in the role played by the web services and the DSS. The web services could take care of integration requirements but for ensuring complete interoperability, a decision model is necessary. In essence, both middleware as well as DSS is required for intelligent buildings. Additionally, a DSS can be of different types which are detailed in Table 8.4.

Intelligent building subsystems produce events which correspond exclusively to their functions. Each event triggered must be conveyed by means of a message which is encoded in a common format in the network. Also, there is a necessity to ensure the best structure for storing the data in databases. This is where a standard message format is the need of the hour

Table 8.4 Types of DSS

Data-based	*Model-based*	*Knowledge-based*	*Document-based*
Data analytics and information systems, storage systems and data bases	Optimization models, empirical models, statistical models	Rules and regulations, facts and procedures etc.	Managing webpages and unfiltered documents

for event sensing, perceiving, and triggering action for interoperability. SOAP (Simple Object Access Protocol), in recent times, has emerged as a popular messaging mechanism between building subsystems. SOAP-based Industrial Messaging Specification (SIMS) entails services where the energy management system and power generation data can be exchanged in real-time. It is highly flexible and can be easily implemented. It has also been used in SCADA (Supervisory control and data acquisition) for tele controlling a very high-capacity power plant comprising of hundreds of hydraulic and tele-controlling units. SOAP makes available an open vendor independent standard for end-to-end communication and high flexibility. This should also be supported by a database module architecture shown in Figure 8.4, for management of the queries of SOAP messages received from the building heterogeneous systems. A SQL-based building management server acts as a mediator for exchange of messages between different subsystems. The module is also responsible for data storage from all the building subsystems along with their related information like status, response, and action codes. Rules also form an integral part of middleware in an intelligent building. Rules enforce subsystems to coordinate and work together. Event–Condition–Action (ECA)-based interoperability framework has been popularly used for intelligent building energy management along with the ability to provide decision support for heterogeneous building subsystems [7–8].

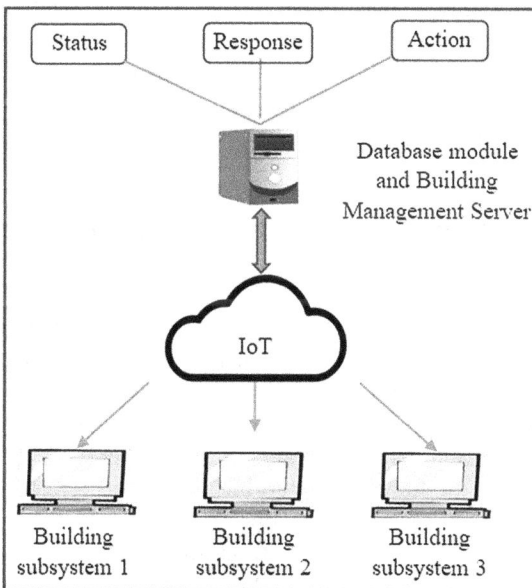

Figure 8.4 Basic architecture of database module.

It is based on Event–Condition–Action (ECA) rules and in the format,

"on" some event
"if" some condition
"then" some actions

The below example elucidates a simple ECA-based rule for interoperation of building subsystems:

On <FA_alarm_triggered>
If <status_zone001_enable>
Do <trigger_PA001>

The ECA rule-based approach revolutionizes the control and management of heterogeneous systems in an intelligent building. Their utility is multifold for the following reasons:

1. *Subsystems are always event-driven and therefore need to be reactive*
2. *Subsystem reaction to an event is also subject to some conditions*
3. *The application logic must be isolated from the execution of procedures*

The ECA interoperation schema defines the rule by employing three-tiered tables, namely the Status, Response and Action tables. The Status table is responsible for ensuring data on subsystems status, internal and external events. The Response table ensures data on subsystems variable and reference codes. Finally, the Action table takes care of Zone ID, action codes, filename, and messages. The subsystems will have their own dedicated action tables configured in the middleware architecture. Also, there should be an element of flexibility in the rules to ensure the facility owners can modify the same [9–10].

8.4 OPPORTUNITIES AND CHALLENGES FOR BEMS INTEROPERABILITY

Interoperable BEMS enhances the building performance while simultaneously reducing the costs. The concept of interoperability in BEMS is presently at the application level after having matured at the development stage. One inherent disadvantage is the cost, which is comparatively more than the existing legacy systems (which are non-interoperable). The installation and commissioning of new interoperable BEMS or retrofitting the existing legacy systems is cumbersome, but the benefits incurred are worthy enough and compensates for any shortcomings. Table 8.5 portrays the benefits accrued from an interoperable BEMS. However, interoperability is also a double-edged sword. There are inherent disadvantages and limitations

Table 8.5 Benefits of interoperable BEMS

Impact of building performance	
Maintenance	Building managers will have more ways to operate and manage facilities. Operations could be consolidated.
Flexibility and operating costs	There is flexibility for the manager to approach any vendor who supports the standards for the features to be implemented. Savings due to vendor independence
Scheduling	A predictive/preventive maintenance program reduces maintenance costs and equipment.
Data access	Huge amounts of information are provided to the managers by the BEMS' interoperable systems, with their seamless access to information across all functions.
Deregulation	Interoperable systems enable managers to control the electrical energy use in real time

present which need to be investigated and rectified for more scalability. The following points need to be addressed:

1. Innumerable internet-enabled devices (create issues for service providers to manage the fault, performance, and security of the devices)
2. Since there is no common standard for IoT security, a high level of privacy and security for users cannot be guaranteed
3. Massive amounts of data need to be stored and processed, leading to higher power consumption.
4. Cloud services, although may serve the purpose of data saving and processing, suffer from standardization and synchronization issues
5. Power requirements for sensing, computing and communication must be optimized as more and more smart devices are integrated
6. Cost and size of the smart devices (with sensing, computing, and communication blocks) must also be optimized.

8.5 CONCLUSION AND FUTURE SCOPE

The chapter presented an overview of the coordination issues faced in BEMS subsystems owing to their heterogeneous nature and how interoperability tries to address the same. It also tries to emphasize the need for a decision support model in the measurement and management of energy usage in buildings. The overall framework of DSS for buildings was also elucidated. Also, the popularly used event driven mechanisms like the Event–Condition–Action (ECA)-based interoperability framework as well as the middleware employed for the same in smart and intelligent buildings is also dealt upon. Since it is an area of continuing research, prominent questions also are raised which would enable researchers in the domain to pander to and find solutions for the same.

REFERENCES

[1] Khajenasiri, I., Estebsari, A., Verhelst, M., Gielen, G. A review on internet of things solutions for intelligent energy control in buildings for smart city applications. *Energy Procedia*. 2017, 1(111): 770–779.

[2] Perumal, T. Making buildings smarter and energy-efficient—using the internet of things platform. *IEEE Consumer Electronics Magazine*. 2021, 10(3): 34–41.

[3] Almusaylim, Z.A., Zaman, N. A review on smart home present state and challenges: linked to context-awareness internet of things (IoT). *Wireless Networks*. 2019, 25(6): 3193–3204.

[4] Mohamed, R., Perumal, T., Sulaiman, M.N., Mustapha, N. Multi resident complex activity recognition in smart home: a literature review. *International Journal of Smart Home*. 2017, 11(6): 21–32.

[5] Power, D.J., Power Enterprises. (n.d.). Ask Dan! about DSS, from http://dss resources.com/faq/index.php?action=artikel&id=13

[6] Le, D.N., Le Tuan, L., Tuan, M.N. Smart-building management system: an Internet-of-Things (IoT) application business model in Vietnam. *Technological Forecasting and Social Change*. 2019, 1(141): 22–35.

[7] Lillo, P., Mainetti, L., Mighali, V., Patrono, L., Rametta, P. An ECA-based semantic architecture for IoT building automation systems. *Journal of Communications Software and Systems*. 2016, 12(1): 24–33.

[8] Park, J.H., Salim, M.M., Jo, J.H., Sicato, J.C., Rathore, S., Park, J.H. CIoT-Net: a scalable cognitive IoT based smart city network architecture. *Human-centric Computing and Information Sciences*. 2019, 9(1): 1–20.

[9] 9 Ways to Create an Intelligent Office. (n.d.). Retrieved January 28, 2021, from https://www.opensourcedworkplace.com/news/9-ways-to-create-an-intelligent-office

[10] Interoperability: 5 Ways It Pays. (2002, August 1). Retrieved January 28, 2021, from https://www.facilitiesnet.com/buildingautomation/article/Interoperability-5-Ways-It-Pays--1710

Chapter 9

IoT-based parking system using Web-App

Samyak Jain, Ankit Gupta and Abhinav Sharma
University of Petroleum & Energy Studies, Dehradun, India

Vinay Chowdary
uGDX Institute of Technology, Hyderabad, India

CONTENTS

9.1 INTRODUCTION

In today's world, there are around one billion automobiles, owning a car is more of a need than a luxury, which highlights the importance of parking places. It is critical to have an adequate number of parking places in cities, malls, offices, schools, hospitals, and other locations in order to accommodate tourists and inhabitants while also avoiding traffic congestion. There is a high density of vehicles, specially in metropolitan areas and in hill stations, that have typically irritated drivers while they waste their time and effort looking for a parking spot and who typically end up parking their vehicles on the street. Further, the rapid growth in the number of vehicles on the road is exacerbating the problem of parking space scarcity. According to industry estimates, around 30% of traffic is caused by tha motorists' failure to find a parking spot. As a result, an effective and smart parking system will save a lot of time, energy, and fuel.

In recent years, the parking management problem has been studied from a variety of different perspectives. With the incorporation of advanced technologies such as the Internet of Things (IoT) [1], artificial intelligence and machine learning [2], smart parking systems have been implemented in developed nations such as the United States, Japan, and so on. IoT [3] is defined as a collection of smart objects (sensors) which can interact with each other through a communication network. In [4] the authors developed a low-cost IoT-based parking system for smart cities. In [5, 6] the authors developed a cloud-based smart parking solution for large parking lots. In [7] the authors utilized the IoT and a genetic algorithm to develop navigation and reservation-based parking system. In [8] authors placed TTGO-ESP32-LoRa boards with ultrasonic sensors to identify free parking slots and then shared this information through the WiFi module present at the end user. In [9] the authors proposed the introduction of a graph-based smart parking system using IoT with its analysis being conducted through a VDM-SL toolbox.

In this contribution the authors have developed an IoT-based smart parking system using low-cost Light Dependent Resistor (LDR) sensors. Today most smart parking systems are based on ultra-sonic sensors or IR sensors [10], but they are costlier and therefore LDR sensors have been used in this work as it is cost-effective and can be deployed easily. The LDR sensor works by the shadow detection method where, with the presence of light or luminous source such as parking lights, luminous intensity is being calculated and with the absence of luminous intensity shadow is being created. Therefore, in this work the LDR sensor has been deployed beneath a vehicle to identify its presence and to indicate the unavailability of a parking slot. XAMPP is used as our local server in the proposed work to test clients or websites before transferring them to a remote web server. The chapter is organized as follows; section 9.2 presents the hardware used and the methodology of this research work; section 9.3 discusses the results; and section 9.4 summarizes the chapter.

9.2 MATERIAL AND METHODS

In this section all of the hardware used to build the IoT-based smart parking system, architecture, step-wise procedure and algorithm is discussed briefly:

a. **Hardware components utilized to build the system**
   i. **LDR Sensor**
      The LDR, also referred to as the photo resistor, photocell is a light-dependent resistor whose resistance changes with the intensity of light impinging on its surface, as shown in Figure 9.1. This resistance finds application in electronic circuits where the detection of light is required. These resistors have range of shapes and sizes with the resistance ranging from kilo-ohm to mega-ohm.
   ii. **Servo Motor**
      A servo motor is a close-packed component that comprises of a DC motor, an integrated circuit (IC), a gear train and a potentiometer, with a bearing for the output shaft, as shown in Figure 9.2. Servos are commonly used in radio-controlled models, such as vehicles, robots, puppets, etc. The servo motor uses a pulse width modulation (PWM) signal to control the DC motor; unlike the

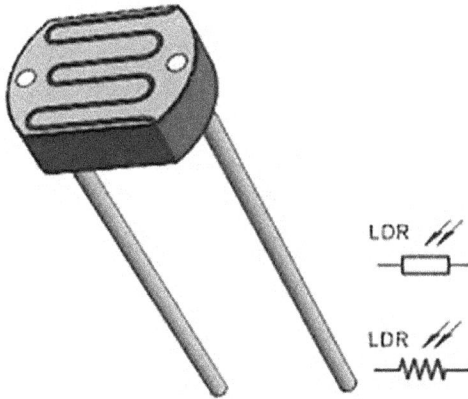

Figure 9.1 LDR sensor.

Figure 9.2 Servo motor.

typical PWM used in conventional DC motors, this PWM signal is utilized to regulate the motor direction or position rather than the rotation speed. Most servo motors function effectively with a PWM frequency of 50Hz, which change the angular rotation from 0 to 180 degrees.

iii. **PIR**

PIRs is a round metal container built from a pyroelectric sensor that consists of a rectangular crystal in the middle which detects infrared radiations, as shown in Figure 9.3.

iv. **NodeMCU ESP8266**

The NodeMCU ESP8266 is a Lua-based open-source firmware development board, as shown in Figure 9.4. It comprises

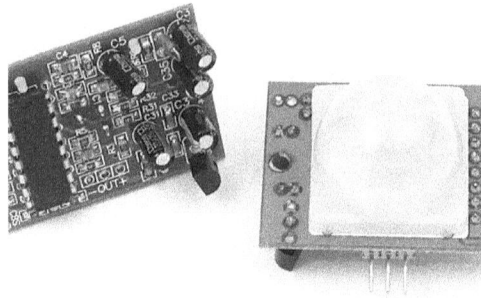

Figure 9.3 PIR sensor.

Figure 9.4 NodeMCU ESP8266.

firmware that operates on Espressif Systems' ESP8266 Wi-Fi SoC, as well as hardware based on the ESP-12 module which contains an ESP8266 chip with a Tensilica Xtensa 32-bit LX106 RISC microprocessor.

v. **Arduino UNO**

The Arduino/Genuino Uno microcontroller board is based on the ATmega328P microcontroller, as shown in Figure 9.5. It comprises of fourteen digital input/output pins, six analogue inputs, a 16 MHz quartz crystal oscillator, a USB connection, a power jack, an ICSP header, and a reset button on the board.

vi. **Ultrasonic (HR-SR04) Sensor**

The ultrasonic sensor, as shown in Figure 9.6, produces a signal with a high frequency. These signals travel with the speed of sound and bounce back to the module if it encounters an obstruction within its path. The sensor has a multi vibrator on its base, which combines a resonator and a vibrator. An ultrasonic wave is produced by the vibrator and transmitted through the resonator.

b. **Architecture**

The architecture of the proposed IoT-based parking system is shown in Figure 9.7. LDR, passive infrared (PIR) and ultrasonic sensors are

Figure 9.5 Arduino UNO.

Figure 9.6 Ultrasonic sensor.

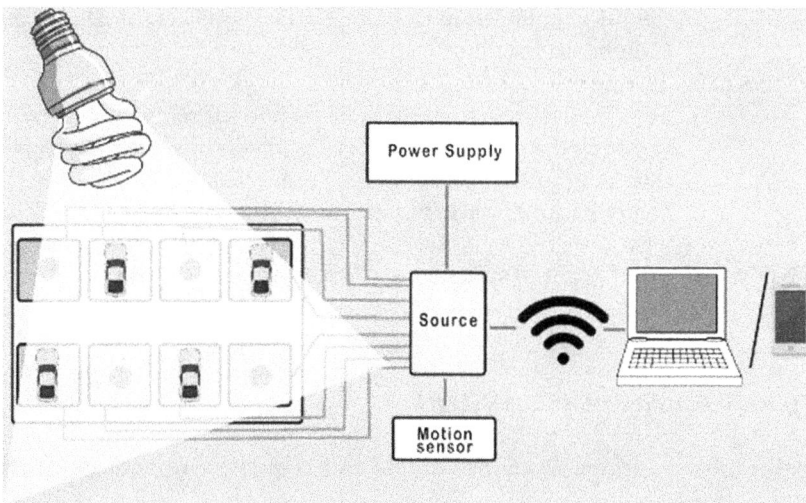

Figure 9.7 Architecture of proposed system for IOT-based smart parking system.

used in this research work. These sensors detect approaching vehicles and find the availability of the parking slots. PIR sensors are used to identify the presence of a car, while an ultrasonic sensor is used at the exit gate to tally how many cars have left the parking lot and LDR is installed on the floor of each parking slot. When a vehicle is parked, it blocks the light. LDR sensor detects light and signals it to Node MCU. Thereafter, based on the received data, NodeMCU communicates with the web server by sending desired sensor values. The web server uses PHP scripts to store and retrieve data from the database server. The web server is accessed using any web app which enables the user to find the free parking slots.

c. **Implementation of the proposed IoT-based parking system**

Step 1: **Setting up the webapp**
Devolve webapp using open-source tools XAMPP to display all the calibrations and calculations as desired outputs on the web application.

Step 2: **Setting up the SQL Tables**
This module will use XAMPP to establish a local SQL server and tables with the needed variables and data types to give us the desired output from the LDR sensor. The real-time sensor values will be stored and viewed using this table.

Step 3: **Setting up the hardware**
This module will integrate LDR, ultrasonic sensor, servo motor, PIR sensor with NodeMCU. Table 9.1 shows the specifications of the components used in the proposed system.

Step 4: **Joining Circuit with Server**
This module will calibrate the sensor values and will provide the Wi-Fi connection to NodeMCU, as well as to the webapp running on localhost.

Step 5: **Testing**
The circuit is tested on the webapp and the required values are stored in the constructed tables by the local server, which will then be graphically displayed on the webapp. Based on this, the empty parking spaces are identified and provided to the user.

The flow chart of the admin system and the user system is shown in Figures 9.8 and 9.9.

9.3 RESULT AND DISCUSSION

The proposed parking system is simple and cost-effective and will prove helpful in developing countries like India. It is simple to deploy and operate, with lower maintenance and operational costs. Therefore, with little

Table 9.1 Specifications of supportive components and use in the proposed system

Component	Specification	Role
Node MCU (ESP8266)	Sixteen digital I/O pins One analog input pin 4MB flash memory 64 KB SRAM Small sized Module In-built Wi-Fi/Bluetooth module chip	Act as a MCU server, Providing hardware connectivity and data transmission to a web server.
LDR Sensor (Light Dependent Sensor)	Range – 100 ohm to 10M ohm Diameter – 3–20 mm Max Power – 200 W	This sensor will assist the user in determining which slots are available and which are reserved.
HC-SR04 Ultrasonic Sensor	Power Supply: +5V DC Working current: 15mA Effectual Angle: <15° Ranging Distance: 2400 cm Resolution: 0.3 cm	The ultrasonic sensor helps us determine whether or not a vehicle is approaching parking, and if it is, the LDR value is refreshed and displayed on the web.
Servo Motor	Voltage: 4.8V–6V Speed: 0.14sec/60 degrees Connector type: JR type	The ultrasonic sensor will communicate a value to the NodeMCU, which will then trigger the servo motor to open the gate.
PIR Sensor	Input Voltage: DC 4.5V~20V Static Current: <50uA Output Voltage: 0V/3V Sensing Range: 7 Meter (120 degree once)	This sensor will assist in detecting the motion of an approaching car, and the servo motor will react in accordance with the sensor's response.
Arduino UNO	Operating Voltage: 5 V Input Voltage: 6V–20 V Digital I/O pins: 14 Analog input pins: 8	Acts as an MCU that interfaces with servo motors, PIR sensors, servo motor, and ultrasonic sensors, as well as providing 5v power to these sensors.

expenses the proposed model can easily be deployed to manage parking problems in metropolitan cities. This model can cut down vehicle time, fuel consumption, and pollutants and can save the time of common people across the country.

To reduce the amount of manual labour, a servo motor is added managed by NodeMCU which controls the automatic opening and closing of parking gates. As shown in Figures 9.10 and 9.11, a webapp is created for both the user and the administrator. The user side application will assist users in determining which slots are booked and which slots are available, while the administrator will be able to monitor the value of the sensor, calibrate ON/OFF, and can access the web application to start and stop displaying the sensor values.

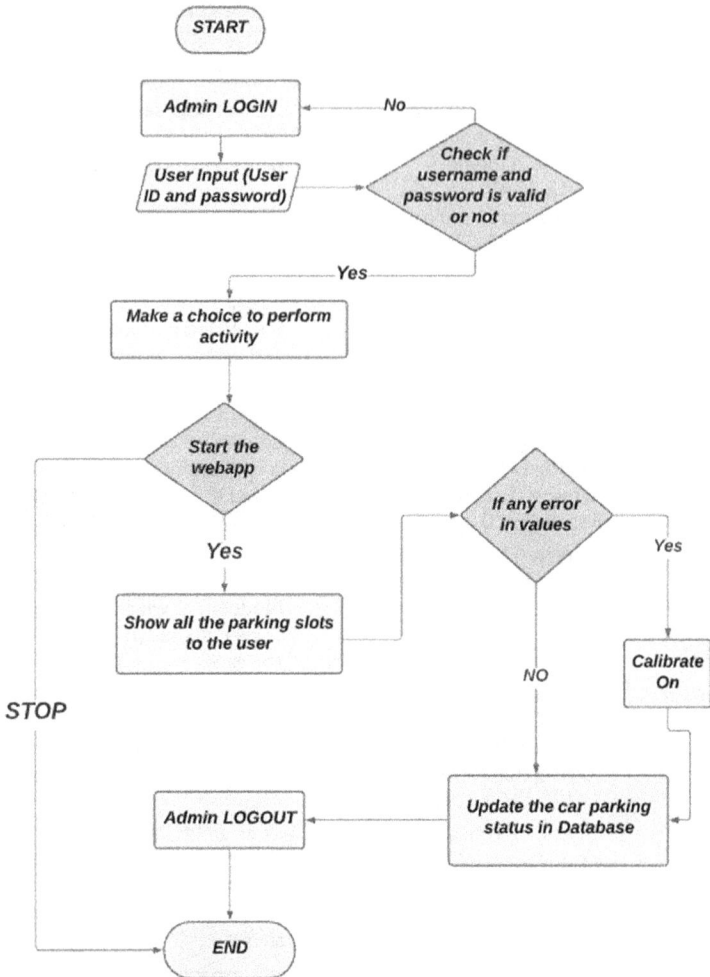

Figure 9.8 Flow chart of admin system.

An XAMPP server shown in Figure 9.12 assists in the setup of the local host and MySQL query in myAdminPHP, which we used to sort the sensor values directly driven by NodeMCU, and the application then queries those values and displays to the user which slot is free and which slot is booked.

A basic prototype with eight slots and LDR sensors installed at the bottom of each slot, which are connected to the NodeMCU as shown in Figure 9.13. The NodeMCU takes the value and sends it to the server database using Wi-Fi, and the web page shows which slot is occupied and which slot is free based on the sensor values.

At the entry gate, a PIR sensor and servo motor, and at the exit gate, an ultrasonic sensor is installed as shown in Figure 9.14. PIR will detect the

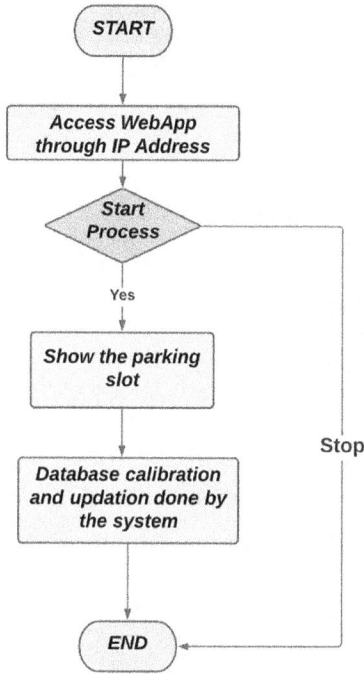

Figure 9.9 Flow chart of user system.

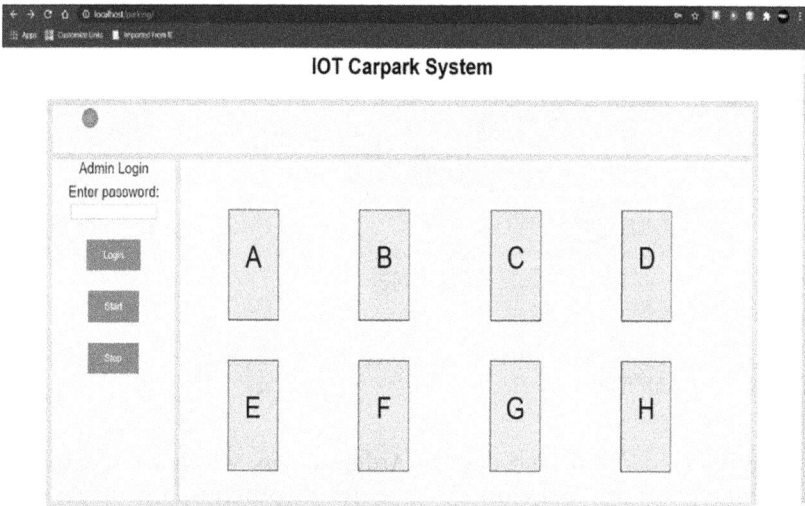

Figure 9.10 User side of web-application.

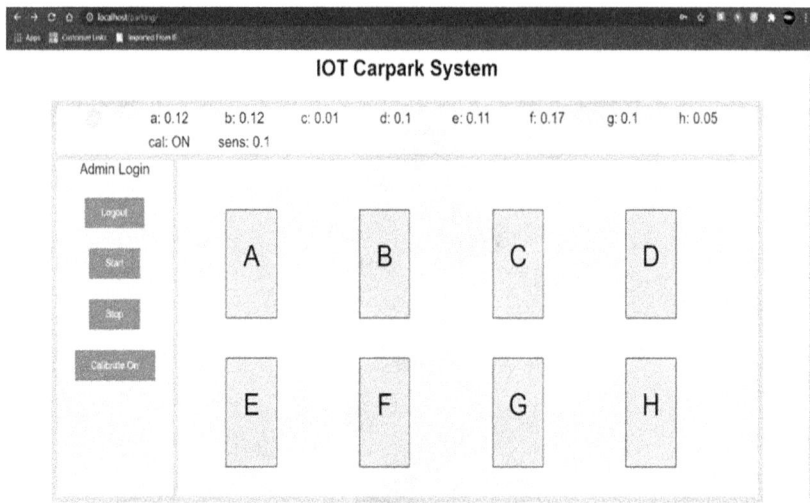

Figure 9.11 Admin side of web-application.

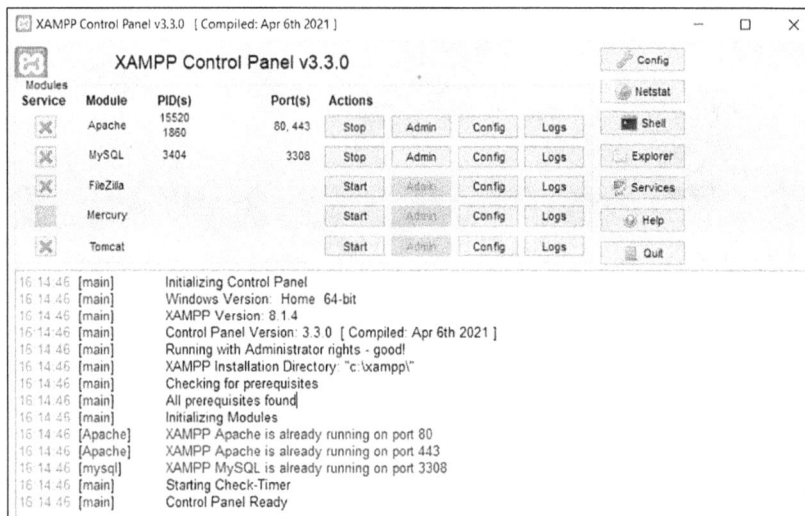

Figure 9.12 XAMPP server.

motion of the car and send data to Arduino Uno, which will control the servo motor to open and close the gate. As we progress, a limit has to be based on the number of parking slots available. For example, if four parking slots are available, the gate will only open four times. At the exit gate, as the car leaves, the ultrasonic sensor will inform the Arduino that another slot is available, and Arduino will send data to the servo motor.

Figure 9.13 LDR connected with NodeMCU.

Figure 9.14 Complete model.

9.4 CONCLUSION

The objective of the proposed research work is to create an internet-based system that allows users to check the availability of parking spaces. With the use of simple and cost-effective components this objective has been achieved. Low-cost sensors, real-time data collecting, and a cloud-based automatic parking system are all part of the proposed system. There are numerous technological alternatives available to ease urban traffic congestion and to improve the efficiency and management of on-street parking. As a scope of future work, deep learning algorithms can be integrated with an IoT-based system to build a smart parking system.

REFERENCES

[1] Xia, F., Yang, L. T., Wang, L., & Vinel, A. (2012). Internet of things. *International Journal of Communication Systems*, 25(9), 1101.

[2] Sharma, A., Jain, A., Gupta, P., & Chowdary, V. (2020). Machine learning applications for precision agriculture: A comprehensive review. *IEEE Access*, 9, 4843–4873.

[3] Chowdary, V., Bera, T., & Jain, A. (2022). IoT solution reference architectures. In *Internet of things* (pp. 39–52). CRC Press.

[4] Vakula, D., & Kolli, Y. K. (2017). Low cost smart parking system for smart cities. In *2017 International Conference on Intelligent Sustainable Systems (ICISS)* (pp. 280–284). IEEE.

[5] Ramasamy, M., Solanki, S. G., Natarajan, E., & Keat, T. M. (2018). IoT based smart parking system for large parking lot. In *2018 IEEE 4th International Symposium in Robotics and Manufacturing Automation (ROMA)* (pp. 1–4). IEEE.

[6] Pham, T. N., Tsai, M. F., Nguyen, D. B., Dow, C. R., & Deng, D. J. (2015). A cloud-based smart-parking system based on Internet-of-Things technologies. *IEEE Access*, 3, 1581–1591.

[7] Aydin, I., Karakose, M., & Karakose, E. (2017). A navigation and reservation based smart parking platform using genetic optimization for smart cities. In *2017 5th International Istanbul Smart Grid and Cities Congress and Fair (ICSG)* (pp. 120–124). IEEE.

[8] Kodali, R. K., Borra, K. Y., GN, S. S., & Domma, H. J. (2018). An IoT based smart parking system using LoRa. In *2018 International Conference on Cyber-enabled Distributed Computing and Knowledge Discovery (CyberC)* (pp. 151–1513). IEEE.

[9] Latif, S., Afzaal, H., & Zafar, N. A. (2018, December). Modelling of graph-based smart parking system using internet of things. In *2018 International Conference on Frontiers of Information Technology (FIT)* (pp. 7–12). IEEE.

[10] Park, Wan-Joo, et al. Parking space detection using ultrasonic sensor in parking assistance system. In *2008 IEEE Intelligent Vehicles Symposium IEEE*, 2008.

Chapter 10

A next-gen IoT-based semi-automatic mobile manipulator

Mukul Kumar Gupta
University of Petroleum and Energy Studies, Dehradun, India

Shival Dubey
Institute of Design, Robotics & Optimization, University of Leeds, United Kingdom

CONTENTS

10.1 INTRODUCTION

The Internet of Things (IoT) has been one of the most familiar features of recent years, scaling new heights and creating a benchmark in the world. It has transformed the real world into an intelligent communication world. In addition, existing challenges are highlighted in these areas.

The value of IoT is increasing, in both commercial and daily settings. In many ways, it is currently making our lives easier and it will certainly continue to do so. It is addressing challenges which we didn't even understand were a challenge, along with the problems we have, before the solution has magically emerged. The chances are that you already understand how useful

DOI: 10.1201/9781003407300-10

IoT is in your daily life through the use of smart thermostats, remote door locks, wired home hubs and all of the different app-controlled appliances.

The application considered in this chapter is a semi-automatic mobile manipulator. A mobile robotic manipulator system is a robotic articulated arm mounted over a mobile base to improve mobility and achieve greater degrees of freedom [1–3]. In such mechanical structures, the degrees of freedom of the robotic arm is enhanced with the degrees of freedom of the mobile base. There is also an improvement in the workspace of the robotic manipulator when it is mounted on a mobile base rather than a fixed base. Any object in free space is represented with a degree of 6 DOF, that is, 3 to represent the object in the cartesian plane and 3 degrees to denote the object orientation. Therefore, to fully manipulate an object in a free space 6 joints are needed to achieve complete movement [4, 5].

When a robotic manipulator has additional degrees of freedom, this is known as kinematic redundancy. When the number of degrees of freedom or robotic system joints exceeds the number of controllable variables, redundancy arises. A kinematically redundant mobile manipulator has more DOFs than is necessary to undertake its task. In this scenario, the inverse kinematics issue yields an unlimited number of solutions. Secondary objectives, such as avoiding mutual limits, singularities, and obstructions, can be better met by selecting mobile manipulator configurations and a motion trajectory from these redundant options.

10.1.1 Classification of mobile robots

Mobile robots can be classified into various categories depending on their locomotion systems, design aspects, and medium for the robot movement and other technical aspects, such as terrain conditions, stability, controllability, manoeuvrability etc. They can be classified into major categories, as illustrated in Figure 10.1, which is based on their locomotion, including stationary robots (arm/manipulators) and land robots. These are then further classified into Wheeled Mobile Robots (WMRs), Legged Robots, tracked locomotion, Hybrid, Aerial robots, water-based robots, space robots, and Bomb Disposal robots [6–9]. WMRs are more energy-efficient than legged robots.

With regard to stability and controllability, both wheel design and wheel geometry play an important role. The stability of a mobile robot is based on the number of wheels used for its motion. The controllability of a mobile robot is the inverse of its manoeuvrability. Coordination between the robotic arm and the platform, which is a speciality of mobile manipulation, is required in various cases for systems combining manipulation and mobility capabilities.

Classification of Manipulators: Manipulators are kinematic chains with special tools as end effectors to handle objects and perform specific tasks, such as welding, assembling and machining operations depending upon

Wheeled Mobile Robot

Under Water Robot

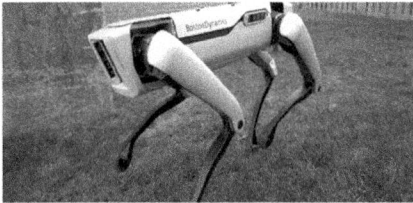

Legged Robot

Aerial Robot

Figure 10.1 Types of mobile robots.

their application. Manipulators can be classified into various criteria based on the following classification:

By Motion Characteristics: Planar manipulator, Spherical manipulator, Spatial manipulator

By Kinematic Structure: Serial robot, parallel robot, Hybrid Robot

10.1.2 Mobile base robotic manipulator

Mobile manipulators are either autonomously operated or teleoperated remotely. A manipulator and a mobile platform are the two subsystems that make up the mobile manipulator system, as shown in Figure 10.2. End effector trajectory and mobile platform trajectory are to be managed simultaneously for the performance of complex tasks such as picking up an object while moving and avoiding an obstacle. Both trajectories should be followed concurrently by the end effector to assure the gripping of the item while avoiding the barrier to align along with the object orientation and the

Figure 10.2 Mobile base robotic manipulator.

mobile base has to follow a trajectory to avoid the obstacle in the case of autonomous systems.

10.1.3 IoT-based robotic manipulator

The number of gadgets connecting with one another has expanded dramatically over recent years. The Internet of Things (IoT) allows items to communicate with one another and to perform tasks such as sensing, identifying, and networking. IoT applications have grown rapidly across all study topics as a result of the limitless number of possible interactions, and they are now leading new research fields.

The Internet of Things is used for remote monitoring and the remote control of devices. Robotic manipulator parameters, such as position, acceleration, tilt, etc., can be monitored to ascertain the manipulator's position and can also be controlled to decide its position. In Figure 10.3, a sample architecture of an IoT-based manipulator is depicted in which a microcontroller acts as the heart of the system. Sensors for measuring important parameters will be placed on the body of the manipulator and the data measured from these sensors will be given to the microcontroller. A predefined algorithm/code will be running on the microcontroller which takes the sensor parameters as an input to decide the position of the manipulator. In order to initiate a control action to the manipulator, a threshold will be set in the algorithm; if the measured parameters breach this threshold then the microcontroller initiates the control action to the manipulator [9–11].

Edge computing is an advanced IoT technique in which information processing is performed on the end device rather than in the cloud. This reduces the computation load on the cloud to a great extent and thus is helpful in fast and parallel processing on the robot side. To implement edge computing on the robot an array of sensors are required, such as:

i. A proximity sensor to provide information on the presence of an obstacle and also to calculate the safe distance between the robot and the obstacle. There are different types of proximity sensors, such as Infrared sensors, Ultrasonic sensors, Passive Infrared sensors (PIR), etc.

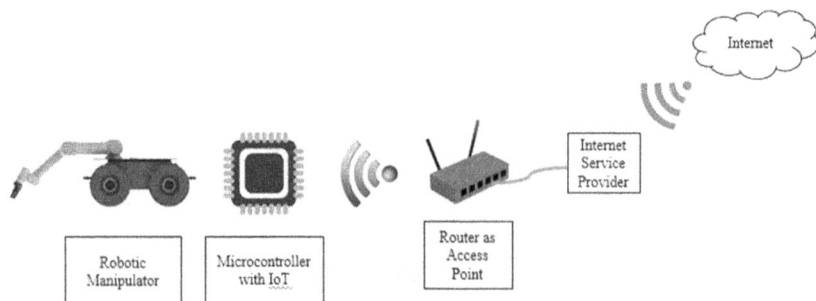

Figure 10.3 IoT based robot manipulator.

ii. A tactile sensor to provide the contact information of a robot and object. As proximity sensors are used to detect objects, tactile sensors are used to avoid objects. Examples of tactile sensors are touch or contact sensors and force sensors, which are used in the grippers of the robot to calculate the force to be applied while gripping any object.

iii. Temperature and humidity sensors to record the operating conditions of the environment where the robot will be operating. Temperature sensors are of contact and non-contact types. Contact-type temperature sensors measure the temperature of liquid materials whereas non-contact temperature sensors operate in the open environment.

iv. Navigation sensors to provide path planning information, along with the position sensor which gives the exact position of the sensor.

v. Acceleration and tilt sensors to provide the rate of robot acceleration and its tilt, if any.

The robot can automatically acquire and send its information and the information about the environment using IoT and sensor technology. With the help of a feedback mechanism on the robot, it is also possible to process this information. One of the key applications based on information from the sensor network is the inspection and control of physical objects with the most up-to-date status, which includes both static properties, viz., the inheritance features of objects such as shape, size, and colour, and dynamic properties, such as the real-time position, movement, gesture, and motion of objects that will change over time, etc. These benefits are critical in allowing large sensor nodes to capture multimodal sensory data in the future ubiquitous IoT framework.

IoT activities are centred on managing, monitoring, and optimizing systems and processes using linked devices with simple, onboard passive sensors. Despite its impact, the IoT solution's full potential could be realized by delving deeper into the more complex and transformative features of pervasive connectivity to and communication among smart objects. Because of their innate ability to detect, think (compute), act (manipulate), and move about, robotic systems can assist in this shift (mobility).

10.2 DESIGN OF THE PROPOSED SEMI-AUTOMATIC ROBOT MANIPULATOR

The proposed mobile manipulator is a 5DOF Manipulator arm with a two-finger gripper as the end effector mounted over a 3 DOF Mobile Robot with a tracked wheel chassis. Altogether this makes it a 9 DOF Mobile manipulator system. The proposed structure is redundant, and it can manipulate in any environment. Redundancy in the manipulator ensures that it can manipulate tasks in an unstructured environment also.

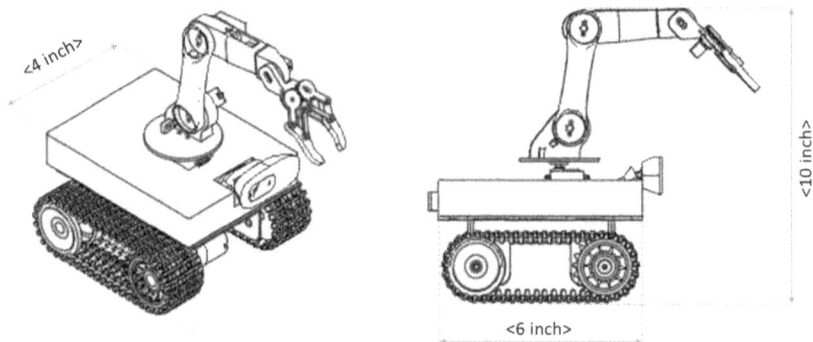

Figure 10.4 Design of the proposed model.

The manipulator considered is depicted in Figure 10.4, with its design features, components attached and dimensions. The drawing is extracted from the 3D model and an isometric view is projected to display sketch dimensions in mm standard units and inches. The proposed design is of the following dimensions: length 6 inches (150 mm), width 4 inches (100mm), and height of 10 inches (254 mm).

10.3 PROPOSED MODEL

The proposed mobile manipulator is of 9 DOF consisting of a 6DOF Manipulator arm with a two-finger gripper as the end effector mounted over a 3 DOF Mobile Robot with a tracked wheel chassis. The 3 DOF mobile robot is an Unmanned ground vehicle (UGV) mounted with a 5 DOF robotic serial manipulator arm on the top with an end effector used to pick and place objects in its workspace. The UGV is also equipped with an FPV camera to obtain the perspective view of the robot to navigate in the environment and also to perform object detection to detect and trace objects around it and identify target objects based on the algorithm. The following solid model, as shown in Figure 10.5, is used to develop mathematical modelling of the system.

Kinematics and dynamics are key parameters in the analysis of robot manipulators. In kinematics under motion, time is not considered directly whereas in dynamics the study of motion considers time. For robot manipulators, there are two approaches for kinematic analysis: the first is forward or direct kinematics, and the second is inverse kinematics. In forwarding kinematics, initial joints are known and the final robot position has to find out whereas in inverse kinematic end-effector positions are known and the robot's initial position has to be found out.

Dynamic analysis can be done either with Euler Lagrange or the Newton Euler approach. With dynamics analysis, one can find out the necessary

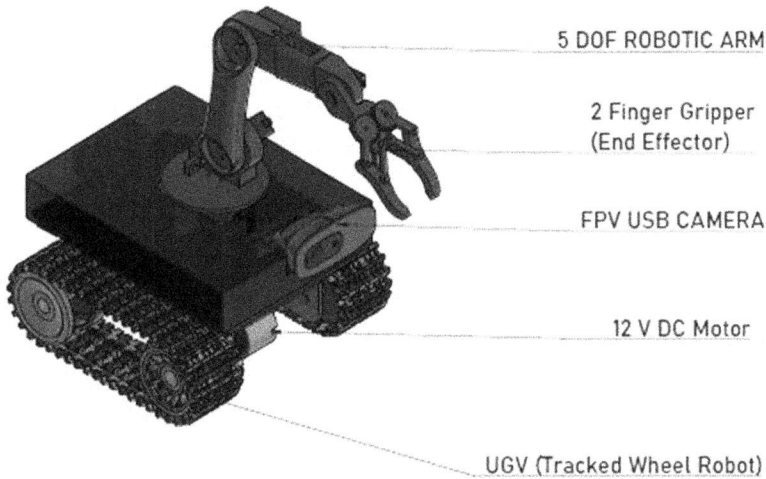

Figure 10.5 CAD model of the proposed mobile manipulator.

torque required for robot deployment. When one has fewer links then the Euler Lagrange approach is better, but in the case of more number links the newton Euler approach is the most suitable.

10.3.1 Denavit-Hartenberg parameters

The Denavit-Hartenberg (D-H) notation is used to describe the location of each link of the manipulator in relation to its neighbouring link. Then, for a given set of joint angles, the resultant pose (position and orientation) is calculated. Any robot can be described kinematically with values of four quantities for each link/joint. Two of these quantities describe the link geometry and the other two describe the link's connection to its neighbouring link.

10.3.2 Frame Assignment for D-H Parameter Calculation

The proposed model uses a 5 DOF Robotic Arm with 5 revolute joints. It consists of a base, shoulder, elbow, wrist and end effector with roll and pitch, as shown in Figure 10.6. To perform the kinematic analysis the standard Denavit-Hartenberg convention and methodology are used to derive the kinematics. This is used to find out the position and orientation of the end effector with respect to its base.

Similarly, 3 virtual frames are considered to write the D-H parameters for the 3 DOF mobile robot base. These frames are corresponding to the three dots of the mobile base, of which two are for translation and one is for rotation (Table 10.1).

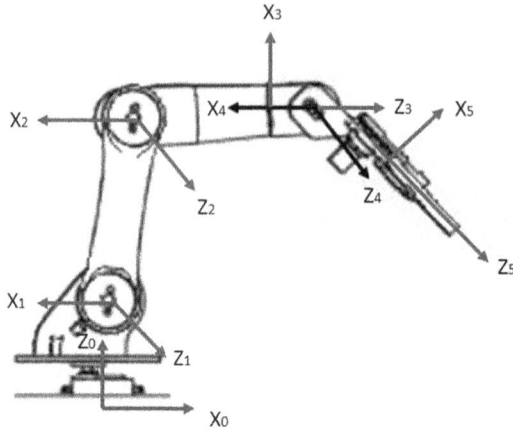

Figure 10.6 Frame diagram for the manipulator.

Table 10.1 D-H parameter of the manipulator

Joint i	Type	α (degrees)	a (mm)	d (mm)	θ_i(degrees)
1	Base	0	0	d_1	θ_1
2	Shoulder	90	0	0	θ_2
3	Elbow	0	a_3	0	θ_3
4	Wrist	90	a_4	0	θ_4
5	Gripper	90	0	d_5	

i	α	a	D	θ
1	3pi/2	0	D1	Pi/2
2	3pi/2	0	D2	3pi/2
3	3pi/2	0	0	θ -3pi/2

10.4 ROBOT CONTROL

The control of robotic manipulators is one of the most challenging tasks as the robot is highly nonlinear in nature. Theoretically, there are various techniques available in nature but in practice mostly PID controller used as a PID controller is easy to implant and cost-effective in nature [12]. The principal nonlinear control techniques are adaptive control, robust control, sliding control, intelligent control, optimal control etc. [13–16]. Stability is also one of the major tasks in the control of robot manipulators as this is one of the main objectives of control. Most of the literature uses the Lyapunov-based approach as this is an energy-based approach. Other theoretical approaches can be divided into joint space-based and task space-based controllers [17, 18].

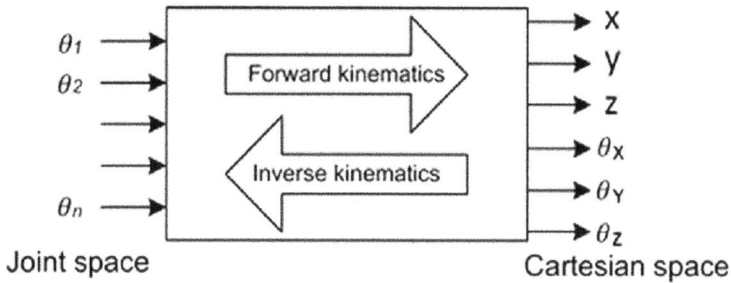

Figure 10.7 Relation between joint space and cartesian space.

Nonlinear dynamics and coupling effects have to consider for a better and more accurate controller design.

10.4.1 Joint space control

The task space is used to train and direct robots to perform certain tasks. In the space-based joint control, inverse kinematic methods are used to map activities into joint space. The relation between joint space and cartesian space is shown in Figure 10.7. The controller is then developed in torque space based on the joint space information. Because the mapping from joint space to task space is injective, the robot may be directed to track the trajectory in joint space, and then the trajectory in task space can be followed. Joint space controllers can be further divided into two groups:

a. classic joint control
b. model-based control

During the early days of robotics, because no computing is required for intricate kinematic and dynamic effects, this solution may be readily conceived, analysed, and implemented using the single-input single-output classical PID technique. Traditional joint control is not only simple to use, but also steady and reliable. This controller is the most popular to date because it is the one most suited to real-world applications. Moreover, recent nonlinear control techniques are far too complex for industrial robotic applications.

10.4.2 Task space control

The task space controller is designed differently from the joint space controller. Task space control has some advantages over joint space control, which are as follows:

• Controllers are designed based directly on task space information so they are more sensitive to environments.

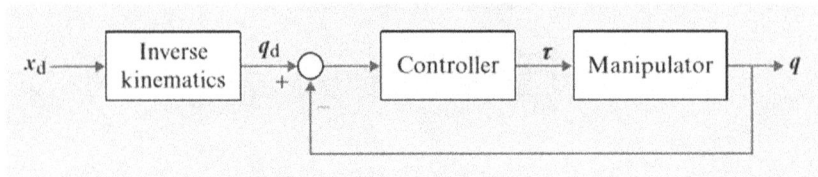

Figure 10.8 Joint space control.

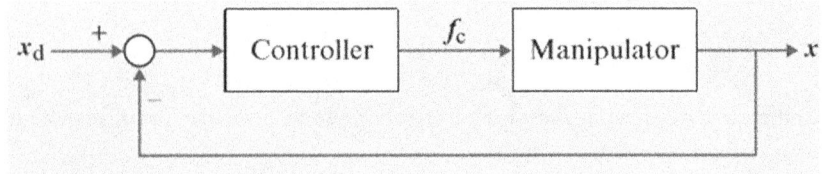

Figure 10.9 Task space control.

- Robots can be properly controlled, even in the case of kinematic singularity.
- Robot programming can be faster.

Task space control, on the other hand, is less well known than joint space control, and its design technique is less clearly defined. In task space, feedback gains are not apparent, and joint space behaviour is difficult to anticipate.

The joint space is the control action whereas the task space is the operational space control.

Figure 10.8 shows the basic outline of the joint space control methods. The main point of the diagram is that torque should be applied to the manipulator and also that one can control position and velocity. The output position should tend to match the desired position as the difference between these two is known as an error. As time tends to infinity, error tends to zero.

Figure 10.9 shows a schematic diagram of the operational space control methods. Because operational space controllers use a feedback loop to directly decrease job mistakes, there are various advantages to this technique. Now, the motion between points can be a straight-line segment in the task space.

10.4.3 Robotic arm simulation using MATLAB GUI

A 5DOF robotic arm is simulated shown in Figure 10.10 using a MathWorks open-source GUI and a robotic system toolbox to derive forward and inverse kinematics. The joint angles are used to determine the cartesian coordinates

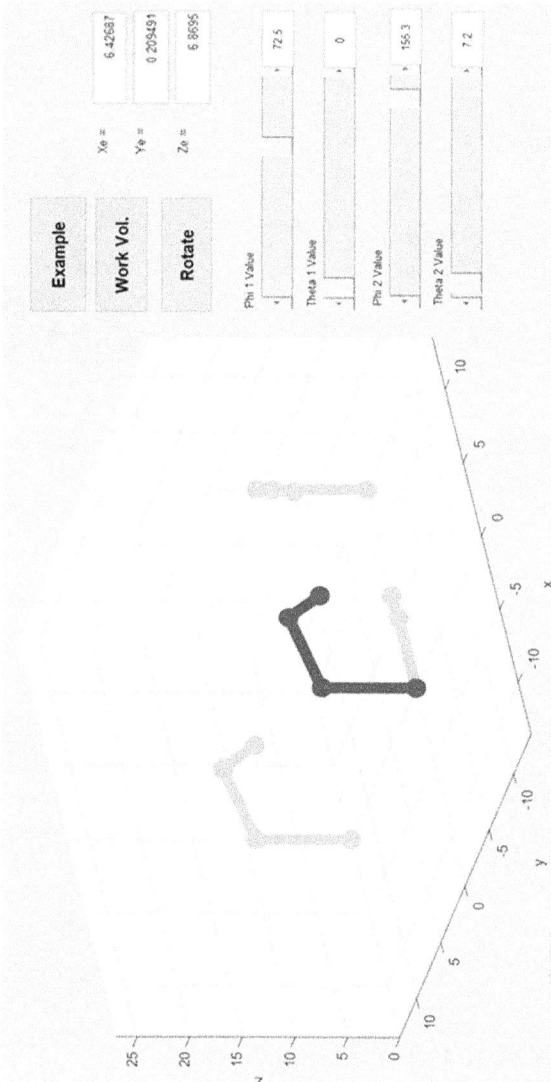

Figure 10.10 3 link manipulator MATLAB GUI simulation.

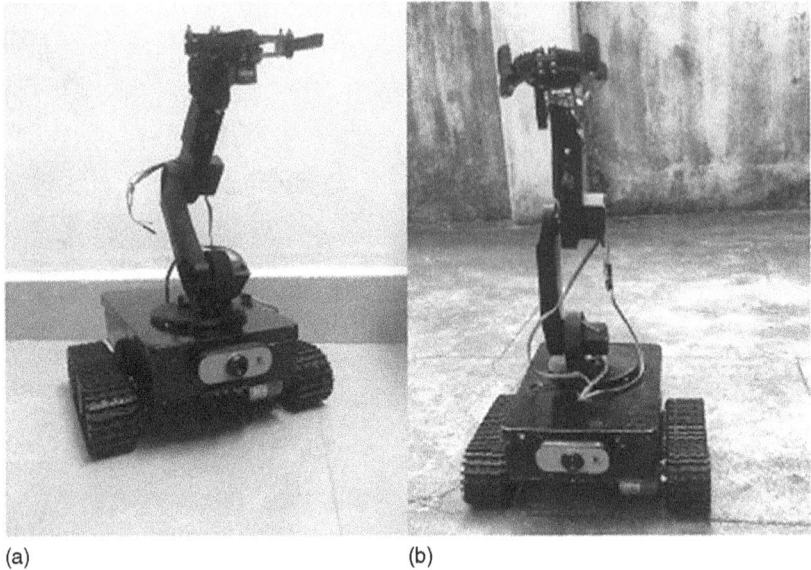

(a)   (b)

Figure 10.11 Robot prototype (a) Indoor environment, (b) Outdoor environment.

of the end-effector and vice versa. The robotic arm when placed on the movable mobile robot base, the end effector coordinates alter with respect to the position of the robot base.

10.5 HARDWARE IMPLEMENTATION

The proposed model, as shown in Figure 10.11, is implemented on hardware using a Raspberry Pi 4 Model B computer and it is teleoperated with the help of the Internet protocol using SSH on a remote laptop or mobile device. The Raspberry Pi computer is programmed using Python scripts to teleoperate the mobile robot using arrow keys and custom designated keys to operate the robotic arm. The Object Detection Algorithm is performed on the USB camera connected to the Raspberry Pi to detect the objects in the robot's field of view [19].

10.6 CONCLUSION AND FUTURE SCOPE

This chapter describes the research that has been carried out to develop a working prototype of a Semi-Automatic Mobile Manipulator using IoT applications. This is developed using a tracked wheel mobile robot. This feature facilitates the robot to navigate conveniently on all platforms irrespective of the terrain as the mechanism is designed to climb rocks and navigate in rugged environments. A 5 DOF robotic manipulator is mounted over the mobile base to improve the workspace of the robotic arm for pick and place applications. The developed system covers mechanics of systems design, dynamic modelling and simulations, and control system design. The mobile manipulator developed is the redundant one. The design of the proposed solid model is done in FUSION 360 software and the proposed model has the manipulator having 5 degrees of freedom on which the mobile base to be mounted has 3 degrees of freedom. For the future scope of this work, the development of a fully autonomous bomb disposal robot with improved object detection algorithms and the addition of an extra manipulator making it a dual-arm mobile manipulator for added dexterity can be implemented. IoT with robotics is mostly used for monitoring the position, the obstacle and other important parameters. It can also be used to control the motion, and the trajectory of the robot along with the implementation of its path planning. The use of IoT is desirable in conditions where human intervention is involved in next to impossible tasks, such as explosive detections, firefighting, forest fire detection and prediction, etc.

REFERENCES

[1] Rubio, F., Valero, F., and Llopis-Albert, C., "A review of mobile robots: Concepts, methods, theoretical framework, and applications". *International Journal of Advanced Robotic Systems*, 16(2), pp. 1–22, 2019.
[2] Ashith Shyam, R. B., Hao, Z., Montanaro, U., Dixit, S., Rathinam, A., Gao, Y., Neumann, G., and Fallah, S., "Autonomous robots for space: Trajectory learning and adaptation using imitation". *Frontiers in Robotics and AI*, 8, pp. 1–10, 2021.
[3] Sivčev, S., Coleman, J., Omerdić, E., Dooly, G., and Toal, D., "Underwater manipulators: A review". *Ocean Engineering*, 163, pp. 431–450, 2018.
[4] Luo, D., and Yu, L., "From factory to site—Designing for industrial robots used in on-site construction" In *Automating cities* (pp. 87–109). Springer, Singapore, 2021.
[5] Van Wynsberghe, A., *Healthcare robots: Ethics, design and implementation*. Routledge, 2016.
[6] Gupta, A., Singh, A., and Verma, V., Amit Kumar Mondal and Mukul Kumar Gupta published in the advanced robotics titled, developments and clinical evaluations of robotic exoskeleton technology for human upper-limb rehabilitation, Vol 34, Taylor and Francis, April 2020.
[7] Asian Defence, "Bomb Disposal Robot Daksh". Theasiandefence.blogspot. com. Retrieved 2010-08-31, 2009.
[8] Naskar, S., Das, S., Seth, A. K., and Nath, A., Application of radio frequency controlled intelligent military robot in defense. In *2011 International Conference on Communication Systems and Network Technologies* (pp. 396–401). IEEE, 2011.
[9] Chowdary, V., Kaundal, V., Sharma, P., and Mondal, A. K., Implantable electronics: Integration of bio-interfaces, devices and sensors. In *Medical big data and internet of medical things* (pp. 55–79). CRC Press, 2018.

[10] Chowdary, V., Mondal, A. K., Sharma, A., and Capoor, S., "Pre-deployment strategy for maximizing barrier coverage in wireless sensor network". *International Journal of Sensors Wireless Communications and Control*, 11(2), pp. 181–188, 2021.

[11] Meera, C. S., Sairam, P. S., Vineeth, V., Adarsh, K., and Gupta, M. K., "Design and analysis of new haptic joysticks for enhancing operational skills in excavator control". *Journal of Mechanical Design by ASME*, 142(12), pp. 460–467, 2022.

[12] Zhou, H., DC servo motor PID control in mobile robots with embedded DSP. In *International Conference on Intelligent Computation Technology and Automation (ICICTA)* (Vol. 1, pp. 332–336). IEEE, 2008.

[13] Epenetus, A. B., Meera, C. S., Mohan, S., and Gupta, M. K., "Development and motion control of spatial serial robotic manipulator under varying disturbances". *World Journal of Engineering*, 16(4), pp. 460–467, 2019.

[14] Gupta, M. K., Kumar, R., Verma, V., and Sharma, A. "Robust control based stability analysis and trajectory of triple link robot manipulator". *Journal Européen des Systèmes Automatisés*, 54(4), pp. 641–647, 2021.

[15] Kanayama, Y., Kimura, Y., Miyazaki, F., and Noguchi, T., "A stable tracking control method for an autonomous mobile robot". In *Proceedings. IEEE International Conference on Robotics and Automation* (pp. 384–389), IEEE, 1990.

[16] Samson, C. and Ait-Abderrahim, K., "Feedback control of a nonholonomic wheeled cart in cartesian space". In *Proceedings. 1991 IEEE International Conference on Robotics and Automation* (pp. 1136–1137). IEEE Computer Society, 1991.

[17] Murray, R. M. and Sastry, S. S., Nonholonomic motion planning: Steering using sinusoids. *IEEE Transactions on Automatic Control*, 38(5), pp. 700–716, 1993.

[18] Khatib, O., "Real-time obstacle avoidance for manipulators and mobile roots". In *Autonomous robot vehicles* (pp. 396–404). Springer, 1986.

[19] Abbaspour, R., "Design and implementation of multi-sensor based autonomous minesweeping robot". In *International congress on ultra-modern telecommunications and control systems* (pp. 443–449). IEEE, 2010.

Chapter 11

Pest identification and classification using IoT enable technique

Atul B. Kathole
D. Y. Patil Institute of Technology, Pune, India

Sonali D. Patil and Kapil N. Vhatkar
Pimpri Chinchwad College of Engineering, Pune, India

Dinesh Chaudhari
JDIET, Yavatmal, India

Avinash P. Jadhav
DRGIT&R Amravati, Yavatmal, India

CONTENTS

11.1 INTRODUCTION

Identifying and classifying crop pests is one of the vital challenges in the field of agriculture [1]. Insects are the most important factor in causing damage to crops and reducing crop productivity. Insect classification is known as a complex task owing to its complex structure and also with the high similarity in appearance among the distinct species [2]. It is very essential for recognizing and classifying the insects present in the crops at an earlier stage by providing highly effective pesticides and also introducing biological control methods to prevent the spread of insects that cause crop diseases [3, 4]. The traditional way of identifying insects is considered to be

DOI: 10.1201/9781003407300-11

inefficient, time-consuming, and labor-intensive. The vision-based computerized system of image processing has been implemented based on machine learning in order to accurately perform the classification and identification of insects so as to overcome the conventional problems in the field of agriculture research.

IoT is a well-known revolutionary technology to enable future communications and computing. The world population is highly dependent on agriculture for income and food resources. Therefore, intelligent Information Technology (IT) technologies are necessary for overcoming the challenges of traditional agricultural approaches. IoT makes it easier for farmers by providing a lot of techniques to achieve sustainable and precise agriculture production in order to address challenges in the agricultural field [5]. IoT technology supports farmers to collect information regarding natural scenarios such as soil fertility, temperature, moisture, and weather. The online crop monitoring system enhances agricultural productivity and crop growth, detects animal intrusion into the field, and is useful in detecting weed, pest, and water levels [6]. IoT enables farmers to monitor their agricultural field anywhere in the world at any time. The main intention of the IoT is to extend the network by concatenating various kinds of connected devices [7, 8]. IoT has mostly focused on three aspects of the system: cost-saving, automation, and communication. IoT encourages people to pursue their routine activities based on the internet and to focus on their principal activity of growing crops [9].

An automated insect identification system is implemented to interpret seven geometrical features. It also makes use of deep learning and machine learning algorithms to obtain better results by considering a smaller number of insect classes. When involving the machine learning algorithms, the classification accuracy has mainly relied on the structure of the extracted features; thus, the optimal features are selected for machine learning, which maximizes the level of accuracy in computation [10]. In addition, the accuracy needs to be improved by incorporating deep learning algorithms for categorizing huge image datasets [11]. The deep learning algorithm is used to perform the automated feature extraction by using the raw data, which decreases the challenges of the hand-crafted features, and also to address the highly complex issues related to image classification. Recently, the deep learning approaches are investigated with the help of Convolutional Neural Networks (CNNs), which ensure promising solutions for the existing challenges [12]. When compared to machine learning techniques, deep learning techniques prove helpful in automatically obtaining the representative features from the training data set, which avoids complex image-processing steps and labor-intensive feature engineering so as to satisfy a wide range of outdoor conditions [13]. An effective deep CNN model is developed for classifying the insect species of field-crop insect images, which provides higher accuracy in classification accuracy. Thus, deep learning techniques have great potential for pest detection in practical applications. Hence, it is significant to develop a new IoT-enabled pest identification and classification model with the help of a deep learning approach.

The main contributions of the research works are given as follows:

- To develop a new IoT-based pest identification and classification model for accurately detecting the pest in the crop field and reducing their effects at the earlier stage for improving the crop production in the agricultural areas.
- To integrate an enhanced deep architecture named CNLSTM for extracting the deep features from the YOLOv3-based detected images and classifying the extracted features along with the parameter optimization using the suggested AHBA to identify the type of pest present in the images.

The remaining section of the developed model is described as follows. Section 11.2 explains the related works and the problems. In section 11.3, the proposed pest identification and classification model is depicted. In section 11.4, the pest identification and classification model is summarized.

11.2 LITERATURE SURVEY

11.2.1 Related works

In 2021, Kumar et al. [14] introduced an enhanced model for identifying the pests that were affecting the rice at the time of crop productivity. Here, the IoT-based mechanism was used for passing the rice pest images to the cloud storage and has provided the pest information. When the pest was identified, the information regarding the presence of the pest was sent to the farmers or owners for taking respective actions. The analysis results have shown that the proposed approach has minimized the rice wastage in the productivity field through the continuous monitoring of the pests in the rice field. In 2020, Chen et al. [15] have implemented a deep learning-based model for obtaining the insect locations and analyzing the environmental information from the weather stations in order to obtain the pests' information in the field with the help of an enhanced deep learning approach. The experimental results have shown that the proposed method have secured improved identification accuracy. Precise identification of the insects and pest has decreased the amount of pesticide usage, which has also minimized the pesticide damage to soil.

In 2021, Turkoglu et al. [16] presented two types of classification approaches with the help of deep feature extraction that were obtained from the pre-trained CNN. The proposed model was validated with the help of diverse diseases and pest images. It was observed that the accuracy scores were better with the majority of the ensemble model and provided improved performance when compared with the existing models. In 2019, Liu et al. [17] have developed an end-to-end method for classifying and detecting the huge multi-class pests with the help of deep learning. The three major parts of the proposed framework were a novel module with an attention-based approach, the developed neural network for ensuring the region proposals, and a score map for classifying the pest and bounding box regression. The experimental analysis was carried out to demonstrate the effectiveness of multi-class pest detection through the proposed model.

In 2021, Li et al. [18] have involved a novel technique for enhancing the accuracy of small pest detection. The suggested framework was trained with the help of a transfer learning methodology using the tiny pest training set. Here, the developed deep learning architecture has provided better performance than other approaches. The analysis has shown that the proposed method has ensured the robust performance in detecting the tiny pests at varied light reflections and pest densities. In 2018, Yue et al. [19] proposed an enhanced residual-based network for detecting the problem in the crop field. The proposed method was correlated with the traditional approaches and has demonstrated the high power capacity of the developed model for image reconstruction. The analysis results have shown that the proposed approach has revealed an enhanced recall rate for pest detection.

In 2021, Wang et al. [20] have developed an efficient deep learning model in the pest monitoring system for automatically detecting and counting the pest in the rice planthoppers. Here, the proposed approach was developed to extract the high-quality regions in the pest images, even the tiny ones. The analysis results have shown that the suggested approach has improved recognition recall compared with the state-of-the-art approaches. In 2019, Thenmozhi et al. [21] implemented an elevated deep learning network for classifying the insect species through the three available data sets. The suggested approach was validated with the other deep learning architectures under the insect classification. Further, this model has included transfer learning for tuning the pre-trained models. The experimental results have shown that the suggested model was effective in classifying the different types of insects and also applicable in the agricultural sector for crop protection.

11.2.2 Problem statement

Plant pests are the most important factor in causing the massive loss in agricultural production, along with the social, ecological, and economical implications. It is essential to recognize and classify the insects present in the crops at the early stage. This is to avoid the insect spread into the crop, resulting in crop diseases by choosing the efficient biological control and pesticides approaches. Numerous features and challenges of agriculture pest detection are reviewed in Table 11.1. Artificial intelligence [14] decreases rice wastage at the time of production by monitoring the pests at the regular

Table 11.1 Features and challenges of agriculture pest detection

Author [citation]	Methodology	Features	Challenges
Kumar et al. [14]	Artificial Intelligence	• It decreases the rice wastage at the time of production through monitoring the pests at the regular interval of time.	• However, the improved technique needs to be developed for better performance.
Chen et al. [15]	YOLOv3, LSTM	• It reduces the damages that are caused in the environment by involving enormous usage of pesticides and also increases the crop quality.	• Yet, there is requirement for enhancing the perspectives in the images to solve the issues related to insufficient training samples.
Turkoglu et al. [16]	Ensemble learning	• It ensures high robustness.	• But, it cannot handle the imbalance problem of training data.
Liu et al. [17]	CNN, Channel-spatial attention	• It is more robust for detecting the tiny pests on an image.	• On the other hand, the size of the sample requires to be maximized in diverse external scenarios for gaining better results.
Li et al. [18]	TPest-RCNN	• It helps to improve the performance by maximizing the replacement frequency of traps.	• However, this model technically difficult because of the limits of computer-vision technology based on visible-range images.

interval of time. However, an improved technique needs to be developed for better performance. YOLOv3, LSTM [15] reduces the damages that are caused in the environment by involving an enormous usage of pesticides and also increases the crop quality. Yet there is a requirement to enhance the perspectives in the images to solve the issues related to insufficient training samples. Ensemble learning [16] ensures a high level of robustness. But it cannot handle the imbalance problem of training data. CNN and channel-spatial attention [17] are more robust in detecting the tiny pests on the image. On the other hand, the sample size requires to be maximized in diverse external scenarios to gain better results. TPest-RCNN [18] helps to improve the performance by maximizing the replacement frequency of traps. However, this model has some technical difficulties because of the limits of computer vision. Deep CNN [19] is used to reduce the density of the monitoring cameras that are employed for surveillance. But there is a need to improve the performance of the model against images with motion blur or noise through the process of DnCNN, Deblur GAN and other image enhancement approaches. CNN [20] performs well with regard to small pest detection. However, this model achieves poor levels of accuracy. CNN [21] has improved potential with regard to pest detection, especially in outdoor applications. But it requires more time for the processing of a large number of data. Therefore, a new pest detection model using deep learning and IoT is required to advance a solution for these abovementioned drawbacks.

11.3 A NOVEL DEEP LEARNING FRAMEWORK FOR IoT-ENABLED PEST IDENTIFICATION AND CLASSIFICATION

11.3.1 Proposed architecture and description

A remote monitoring mechanism with IoT devices is the vital technique involved in diverse applications such as object tracking in smart cities, healthcare, human surveillance, and modern farming. With the help of IoT, insect control can be monitored and managed from anywhere in the world. In [11], the IoT network is used along with a wireless imaging system for developing the remote greenhouse pest monitoring system. Here, the blob counting and k-means color clustering algorithm are used in the imaging system to automatically count the insects and pests present in the trap sheet. Similarly, in another research study, the IoT-based smart farm field management methodology is implemented to monitor crop growth, detecting the insects in the crop field and also determining the appropriate pesticide to manage the crop pests. On the other hand, the automated identification of the insects and pests are the major challenges in the pest monitoring systems. Hence, the deep learning and machine learning algorithms are used for the decision-making and object detection in diverse insect control mechanisms. In [12], the effective performance of the machine learning, deep learning, and computer vision algorithms are utilized for pest detection, especially in the tomato farms. This study has shown that the deep learning architecture provides enhanced performance when compared among three considerations. From the literature, it is clear that IoT is more important for pest monitoring systems, whereas the deep learning techniques are known to be the optimal approach for detecting and classifying insects and pests from the crop images. Thus, IoT and deep learning are combined into the pest identification and classification model in order to offering increased benefits to the farmers. The architecture has been diagrammatically represented in Figure 11.1.

A new pest identification and classification model with IoT using the deep learning architecture is developed to identify and classify the pests in the crops so as to reduce the usage of fertilizers and increase crop production by preventing the pests at an earlier stage of crop growth. The IoT technology

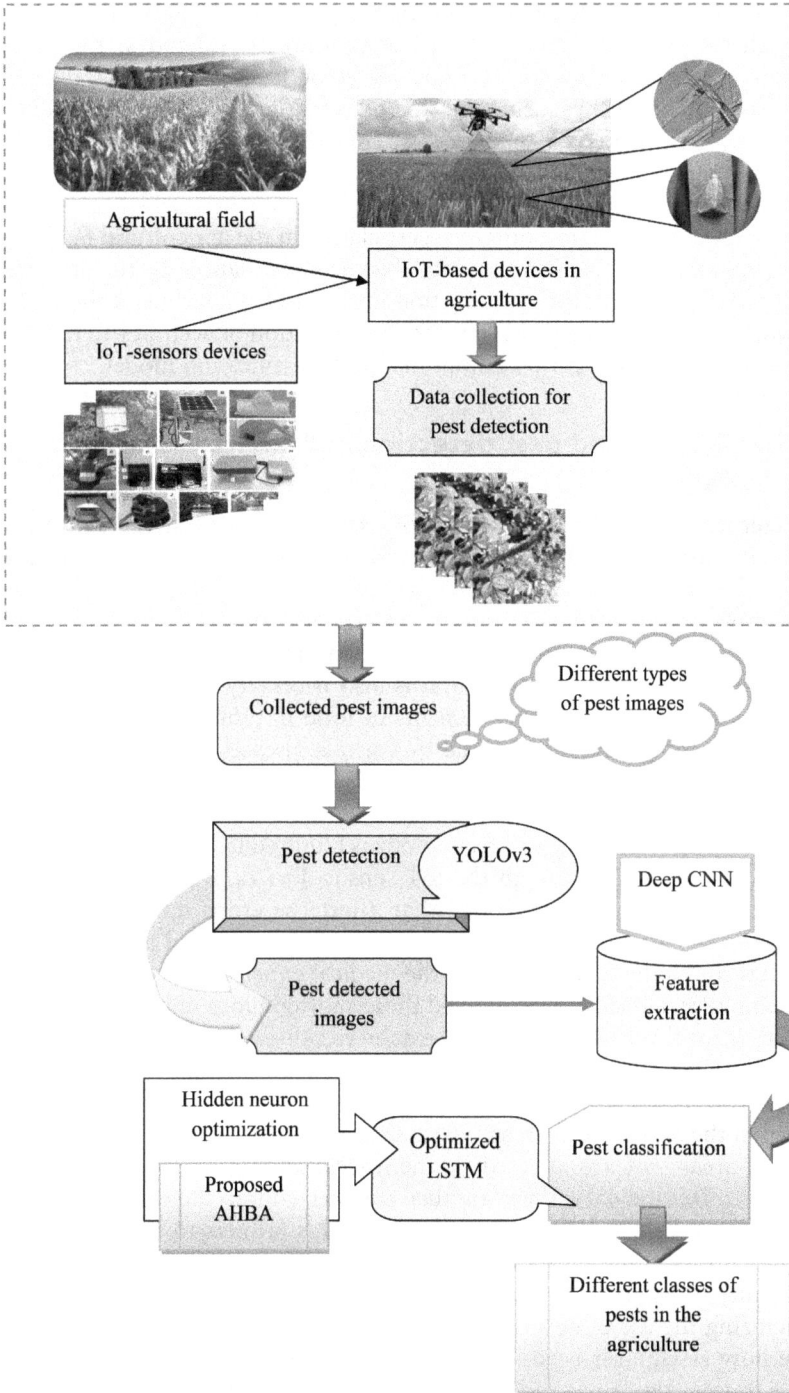

Figure 11.1 Proposed pest identification and classification architecture with IoT and deep learning approach.

is used for collecting the required crop images from the agricultural fields through the sensors. These collected images are considered at the object detection phase, where the YOLOv3 detector is used to detect the pest regions in the given input images. The detected images from the YOLOv3 are given to the CNLSTM network, where the CNN framework is used to extract the most essential features of the detected pest images. The pest features are passed toward the developed LSTM network, which is used to classify these features into different pest classes in the agricultural field. The performance of the classification model is further enhanced by the proposed AHBA on conducting the optimization in the hidden neurons of the LSTM network, which intends to achieve the maximization of accuracy in the classification phase in the proposed optimal pest classification model.

11.3.2 IoT-enabled pest detection and smart agriculture

Agriculture is defined to be a science of crop production, animal husbandry and soil cultivation, where the resultant products need to be marketed in an effective manner. Food demand has been increasing abundantly, in both qualitative and quantitative aspects, which can be satisfied by incorporating the computer technology into the agricultural practices. Owing to the growth in the world's population, it is also necessary to increase the crop production regarding the requirements of food in terms of nutrition. Crop production can be mainly affected by various diseases that are caused due to certain factors, such as the presence of insects and pests in the crop field. This needs to be prevented by reducing the disease-affected crop in the agricultural areas. Hence, the IoT technology is required to automatically detect the insects and pests through the IoT sensor devices, which helps farmers to monitor and remove the pests that affect the crops during the earlier stages of crop growth. In the traditional way, the insects and pests are determined through the manual way by the medical experts, which may be time-consuming and inaccurate. To avoid these shortcomings, the automatic and remote identification of pests can be achieved through IoT. IoT-based agricultural systems provide accurate results when determining the pest regions in the crop field. Here, the sensors need to be properly installed and maintained in the agricultural areas. These sensors help farmers to identify those planted areas that are affected by the insects and pathogens. Initially, the sensors collect the data; these are then transferred into the centralized platforms through the wireless network. This helps farmers to remotely monitor and protect their crops from insect and pest attacks, also reducing the possibility of environmental contamination and crop intoxication and also minimizing the use of pesticides in the crop field. These developments illustrate how essential it is to incorporate IoT technology with agriculture in order to improve crop production. The IoT-based system for pest detection in smart agriculture is shown in Figure 11.2.

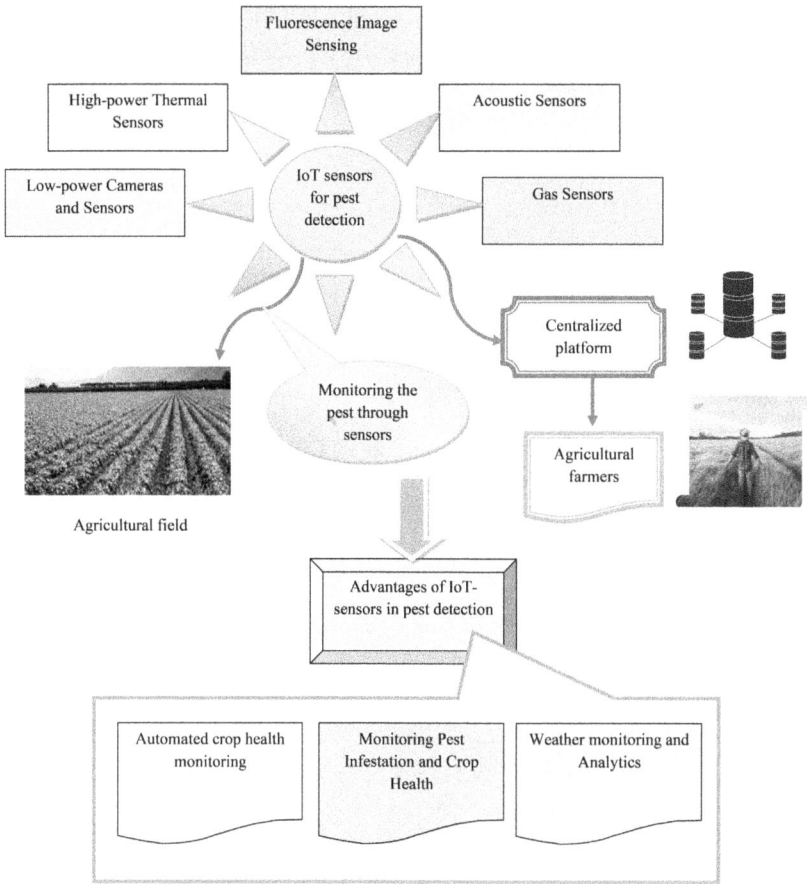

Fluorescence Image
Sensing

High-power Thermal
Sensors

Acoustic Sensors

Low-power Cameras
and Sensors

IoT sensors
for pest
detection

Gas Sensors

Centralized
platform

Monitoring the
pest through
sensors

Agricultural
farmers

Agricultural field

Advantages of IoT-
sensors in pest detection

Automated crop health
monitoring

Monitoring Pest
Infestation and Crop
Health

Weather monitoring and
Analytics

Figure 11.2 IoT-based pest detection in smart agriculture.

11.4 CONCLUSION

This research has proposed a novel pest identification and classification model with implemented AHBA in order to achieve the accurate detection of pest in the crop field. Initially, pest images were collected through sensors by using IoT technology. The collected images were subjected to the object detection phase, during which the YOLOv3 detector was utilized to detect the pest regions in the given input images. The detected images were obtained from the YOLOv3, and these were further given into the CNN framework to extract the significant features from the pest images. The extracted features were considered for the developed CNLSTM network, where the optimal feature classification with proposed AHBA was performed that has classified into different classes of pests in the agricultural field. Thus, the overall performance of the proposed pest identification and classification model with implemented AHBA-CNLSTM was superior to the other conventional techniques. Future research should try to implement the system on real-time images for better performance analysis when compared with existing approaches.

REFERENCES

[1] M. Khanramaki, Ezzatollah Askari Asli-Ardeh and Ehsan Kozegar, "Citrus pests classification using an ensemble of deep learning models," *Computers and Electronics in Agriculture*, Vol. 186, No. 106192, pp. 1–11, July 2021.

[2] M. Turkoglu, D. Hanbay and A. Sengur, "Multi-model LSTM-based convolutional neural networks for detection of apple diseases and pests," *Journal of Ambient Intelligence and Humanized Computing*, pp. 1–11, 2019.

[3] F. Wang, R. Wang, C. Xie, P. Yang and L. Liu, "Fusing multi-scale context-aware information representation for automatic in-field pest detection and recognition," *Computers and Electronics in Agriculture*, Vol. 169, No. 105222, pp. 1–11, February 2020.

[4] Thenmozhi Kasinathan and Srinivasulu Reddy Uyyala, "Machine learning ensemble with image processing for pest identification and classification in field crops," *Neural Computing and Applications*, Vol. 33, pp. 7491–7504, 2021.

[5] M. Ayaz, M. Ammad-Uddin, Z. Sharif, A. Mansour and E.-H. M. Aggoune, "Internet-of-Things (IoT)-Based Smart Agriculture: Toward Making the Fields Talk," *IEEE Access*, Vol. 7, pp. 129551–129583, 2019.

[6] C.-J. Chen, Y.-Y. Huang, Y.-S. Li, Y.-C. Chen, C.-Y. Chang and Y.-M. Huang, "Identification of fruit tree pests with deep learning on embedded drone to achieve accurate pesticide spraying," *IEEE Access*, Vol. 9, pp. 21986–21997, 2021.

[7] Y. Ai, C. Sun, J. Tie and X. Cai, "Research on recognition model of crop diseases and insect pests based on deep learning in harsh environments," *IEEE Access*, Vol. 8, pp. 171686–171693, 2020.

[8] A. B. Kathole and Dr. Dinesh N. Chaudhari, "Pros & Cons of Machine learning and Security Methods," 2019. http://gujaratresearchsociety.in/index.php/JGRS, ISSN: 0374-8588, Vol. 21 N4.

[9] A. B. Kathole, Dr. P. S. Halgaonkar, and A. Nikhade, "Machine learning & its classification techniques," *International Journal of Innovative Technology and Exploring Engineering (IJITEE)* ISSN: 2278-3075, Vol. 8 No. 9S3, pp. 1–7, July 2019.

[10] A. B. Kathole and Dr. D. N. Chaudhari, " Fuel analysis and distance prediction using machine learning," *International Journal on Future Revolution in Computer Science & Communication Engineering*, Vol. 5, No 6, pp. 1–11, 2019.

[11] Mohamed Smail Karar, Fahad Alsunaydi, Sultan Albusaymi and Sultan Alotaibi, "A new mobile application of agricultural pests recognition using deep learning in cloud computing system," *Alexandria Engineering Journal*, Vol. 60, No. 5, pp. 4423–4432, October 2021.

[12] Everton Castelão Tetila, Bruno Brandoli Machado, Gilberto Astolfi, Nícolas Alessandro de Souza Belete, Willian Paraguassu Amorim, Antonia Railda Roel and Hemerson Pistori, "Detection and classification of soybean pests using deep learning with UAV images," *Computers and Electronics in Agriculture*, Vol. 179, No.105836, pp. 1–11, December 2020.

[13] D. J. A. Rustia, Chen-YiLu, Jun-Jee Chao, Ya-Fang Wu, Jui-Yung Chung, Ju-Chun Hsu and Ta-Te Lin, "Online semi-supervised learning applied to an automated insect pest monitoring system," *Biosystems Engineering*, Vol. 208, pp. 28–44, August 2021.

[14] S. K. Bhoi, K. K. Jena, S. K. Panda, H. V. Long, R. Kumar, P. Subbulakshmi and H. B. Jebreen, "An internet of things assisted unmanned aerial vehicle based artificial intelligence model for rice pest detection," *Microprocessors and Microsystems*, Vol. 80, No. 103607, pp. 1–11, February 2021.

[15] C. J. Chen, Y. Y. Huang, Y. S. Li, C.-Y. Chang and Y.-M. Huang, "An AIoT based smart agricultural system for pests detection," *IEEE Access*, Vol. 8, pp. 180750–180761, 2020.

[16] M. Turkoglu, B. Yanikoğlu and D. Hanbay, "PlantDiseaseNet: convolutional neural network ensemble for plant disease and pest detection," *Signal, Image and Video Processing*. Received: 27 August 2020 / Revised: 19 February 2021 / Accepted: 5 April 2021 © The Author(s), under exclusive licence to Springer-Verlag London Ltd., part of Springer Nature 2021, 1–9, https://doi.org/10.1007/s11760-021-01909-2

[17] L. Liu, R. Wang, C. Xie, P. Yang, F. Wang, S. Sudirman and W. Liu, "PestNet: An end-to-end deep learning approach for large-scale multi-class pest detection and classification," *IEEE Access*, Vol. 7, pp. 45301–45312, 2019.

[18] W. Li, D. Wang, M. Li, Y. Gao, W. Jianwei and X. Yang, "Field detection of tiny pests from sticky trap images using deep learning in agricultural greenhouse," *Computers and Electronics in Agriculture*, Vol. 183, No. 106048, pp. 1–11, April 2021.

[19] Y. Yue, X. Cheng, D. Zhang, W. Yunzhi, Y. Zhao, Y. Chen, G. Fan and Y. Zhang, "Deep recursive super resolution network with Laplacian Pyramid for better agricultural pest surveillance and detection," *Computers and Electronics in Agriculture*, Vol. 150, pp. 26–32, July 2018.

[20] F. Wang, R. Wang, C. Xie, J. Zhang, R. Li and L. Liu, "Convolutional neural network based automatic pest monitoring system using hand-held mobile image analysis towards non-site-specific wild environment," *Computers and Electronics in Agriculture*, Vol. 187, No. 106268, pp. 1–10, August 2021.

[21] K. Thenmozhi and U. Srinivasulu Reddy, "Crop pest classification based on deep convolutional neural network and transfer learning," *Computers and Electronics in Agriculture*, Vol. 164, No. 104906, pp. 1–11, September 2019.

Chapter 12

Framework for leveraging diagnostic and vehicle data with emphasis on automotive cybersecurity

Thrilochan Sharma
L. & T. Technology Services Limited, Bangalore, India

K. S. Ashwini
Coventry University, United Kingdom

Prithvi Sekhar Pagala and KNS Acharya
L. & T. Technology Services Limited, Bangalore, India

CONTENTS

12.1 INTRODUCTION

Transportation and automobiles have become a pivotal component for a country's economy and development [1]. Presently, automobile systems are evolving into multiple forms according to the requirements of the contemporary world. In recent years, automotive systems have achieved a level of maturity, progressing from mechanical to electromechanical systems and moving ahead with applications such as connected, autonomous, and electric vehicles. Advances have been observed in applications which have increased the scope of data collection, networks both in the vehicle and outside, with processing and usage for advanced vehicle features [2]. This has increased confidence in the design of automotive systems which have assisted in the creation of autonomous vehicles. The Society of Automotive Engineers (SAE) has drawn up a classification of Level-0 to Level-5 for motor vehicles and their operations on public roads [3]. Within this system, Level-0 is non-autonomous, as can be seen in today's standard, non-connected vehicles. Level-1 and Level-2 autonomous vehicles provide relief to the driver by automating vehicle controls such as the brake, the accelerator, and the steering. The Advanced Driver Assistance System (ADAS) is a standard feature in today's vehicles, which performs advanced operations such as cruise control, automated braking, lane assist, parking assist, and so on [4]. Level-3 and Level-4 autonomy helps a vehicle to run autonomously on standard highways, and restricted environments such as university complexes, golf courses, and industrial parks. Finally, there is Level-5 autonomy where the system takes over all vehicle controls and requires no human intervention while driving. The Level-5 vehicles that are currently being designed do not even have steering or pedals.

The primary objective of the On-board Diagnostics (OBD) system when it was standardized by SAE in 1979 was to improve in-use emission compliance through the monitoring of the computerized emission control system. In 1988, the California Air Resource Board prepared and implemented the successor system, OBD-II. The purpose was to monitor the performance of sensors and actuators affecting the engine emission norms and finally, from 1994 onward, it was applied to all vehicles [5]. Engine monitoring, misfire warning, fuel system analysis, air conditioning system, etc. were the added features to monitor the various sub-systems of a vehicle. Most of the monitoring aspects of the data collected through the OBD port were indicated using Malfunction Indicator Lamps (MIL). India's first official policy, General Statutory Rule-84, was released in 2009, providing guidelines for Bharat Standard IV (BSIV)-compliant vehicles [6]. From 2010, all the four-wheelers in India are mandated by the government to have an OBD port; further, vehicles manufactured from April 2020 onward must be equipped with the OBD system as their standard feature in line with Bharat Standard VI [7].

As the vehicles became increasingly complex, this resulted in a huge rise in the data generation and usage by sensors and actuators. This has increased the scope for research and data inferences, resulting in the introduction of new applications which make use of the collected data. Vehicle parameters, such as acceleration, speed, engine rpm, throttle position and engine load, have served as critical features in the studies for vehicle and, in particular, the monitoring of engine health. The analysis of these parameters supplies information regarding improper gear shifting and speeding. Driving at high speed at lower gear increases the friction loss [8] and speeding increases the load on the engine, which leads to increased fuel consumption [9]. The discussed parameters can also be used to analyze the driver's behavior. With the increase in the abundant amount of data and the inferences that are made from it, every vehicle has become a computer. They can be considered as mobile nodes when compared to a generic IOT network.

Connected vehicles (CVs) is the concept by which the vehicle can communicate bidirectionally with other systems outside of the vehicular systems, including other vehicles and road infrastructure. CVs have already been adopted in the market to enable remote and advanced features for users, along with helping in increasing the safety and quality of vehicle transportation. Research has confirmed that the network of CVs helps to decrease the number of accidents and the avoidance of road congestion [10]. Vehicles will be equipped with the necessary hardware to achieve Vehicle to X (V2X) communication, which typically allows a vehicle to communicate with nodes in its environment, including other vehicles (V2V), the cloud/ network (V2N), the infrastructure (V2I), and pedestrians (V2P) [11]. A study conducted in 2014 had forecasted that the number of sensors in a vehicle would reach a total of 400 by 2020 [12]. Vehicle to Sensor (V2S), V2I, V2N, and V2V systems are studied and wireless efficient solutions are provided by previous studies [13]. Inter-vehicle co-operation channel estimation (IVC-CE) is one of the methods which will enhance the performance and support for channel estimation in the case of V2I applications with a special focus on safety-related applications has been illustrated [14]. Geographically, the North American automotive market currently holds the largest share of the global connected vehicles, followed by the European market [15]. With the advancement of such intelligent vehicles, the focus shifts to sensory networks within the vehicle which can enable efficient communication.

Through network connectivity, such a vehicular network provides abundant amounts of different types of data from the various sensors of all road vehicles. The inclusion of advanced sensors in modern vehicles such as depth cameras, Lidars, Radars, etc., has led to numerous applications using advanced technologies such as image processing, object detection, feature extractions, depth calculation, localization, path planning, etc. Artificial intelligence (AI), machine learning (ML) and deep learning (DL) all play major roles in future when these sensors become standardized in automotive systems. The analysis of health conditions, the prediction of failures, behavioral analysis, and the diagnostics of vehicular systems, etc., are the immediate applications possible through the implementation of AI, ML and DL.

Advancements in automotive features, with increased data flow in a vehicle, also attract the attention of users with negative intentions posing threats and exploiting the vulnerabilities of the network using cyberattacks. The training and testing of ML algorithms for simulating and resolving attack models to achieve secure networks is one of the important steps in the application of AI/ML in CV networks [16]. These scenarios provided a necessity to protect the automotive systems and the data contained in them, giving rise to the need for cybersecurity. Cybersecurity has become highly prominent due to the continuous increase of vehicle network interconnection and intellectualization [17].

Cyber-risks of autonomous, connected vehicles have gained a prominent attention over the past few years as there as various technical vulnerabilities that have uncovered by security researchers. Cybercriminals can disrupt the functionalities of vehicular systems such as steering, braking and the engine by hijacking the Electronic Control Units (ECUs). V2X has opened the possibility of the vast spectrum of applications, but also has a number of risks involved in terms of cyberattacks. Involving multiple devices being paired to the vehicular network, such as a smartphone, network connectivity hardware,

Table 12.1 Overview of major standards in design and development of automotive systems

Standard	First release	Primary goal
ISO 26262	June 2009	Unifying safety standard for all automotive electrical/ electronic systems [17]
ISO/PAS 21448	Jan 2019	Guidance on the applicable design, verification and validation measures needed to achieve the safety of the intended functionality (SOTIF) [18]
ISO/SAE 21434	May 2020	Baseline for vehicle manufacturers and suppliers to ensure that cybersecurity risk management is done efficiently and effectively [19]

etc., provides certain gateways for the cybercriminals. Standards play a major role in the reduction of such risks and maintain uniformity across platforms in terms of automotive software. ISO and the SAE are primarily responsible in the design and maintenance of automotive standards. Components such as cybersecurity risks in road vehicles, functional safety throughout the development cycle of vehicles, etc., are defined as per the standards mentioned in Table 12.1.

These standards are the documents that are taken as reference while automotive electronic systems are being designed, developed, and tested. These standards help the automotive industry to achieve safety and security, resulting in reliable automotive systems.

The drawback of the automotive industry currently is that the market is predominantly occupied by non-connected vehicles which are incapable of achieving features of a connected vehicle and benefiting from the communication network of connected vehicles. The worldwide automotive market was comprised of around 8% CVs in 2018, a figure which is projected to rise to 23% by 2023. The Indian CV market is projected to be US$32.5 billion by 2026 from a value of $9.8 billion in 2021 [20]. Such a scenario creates a number of challenges when a traffic environment is comprised of both non-connected vehicles and connected vehicles. The application of converting a non-connected vehicle into an IOT node with additional hardware remains a challenge and has received little research attention. A framework for conversion of a non-connected vehicles into an IOT node is needed in order to achieve both efficiency and safety.

12.2 FRAMEWORK

The growing population density and traffic conditions demand higher vehicular intelligence in the domain of information and communication technologies. The implementation of the V2X technology through connected vehicles has the potential to make road transport both safer and more reliable by establishing communication with several vehicular and infrastructural entities. In general, the vehicles that provide a data-rich environment that gives rise to multiple transport-related applications, making them the building blocks to achieve V2X communication can be termed as "**AutoNodes**". Each AutoNode can function as an independent network device continuously contributing to the framework. Data obtained from such AutoNodes can be very useful in enhancing multiple data pipelining levels in vehicles, all of which can contribute to providing an information-rich travel experience. Since CVs are equipped with various sensors and data transceiver devices which contribute to the vehicular IOT network, they can be AutoNodes by default, as shown in Figure 12.1. As discussed earlier, however, the major challenge with regard to the emerging automobile market of India is that the number of non-connected vehicles is greater than the CVs. In such cases, non-connected vehicles can be converted into connected vehicles, making them capable of interacting with the proposed framework serving as an AutoNode. Once non-connected vehicles are converted into AutoNodes, they can benefit from the CV network by performing both data collection as well as data reception activities.

In general, the non-connected vehicles' framework requires external sensory equipment to access vehicular ECU data and it can be converted into

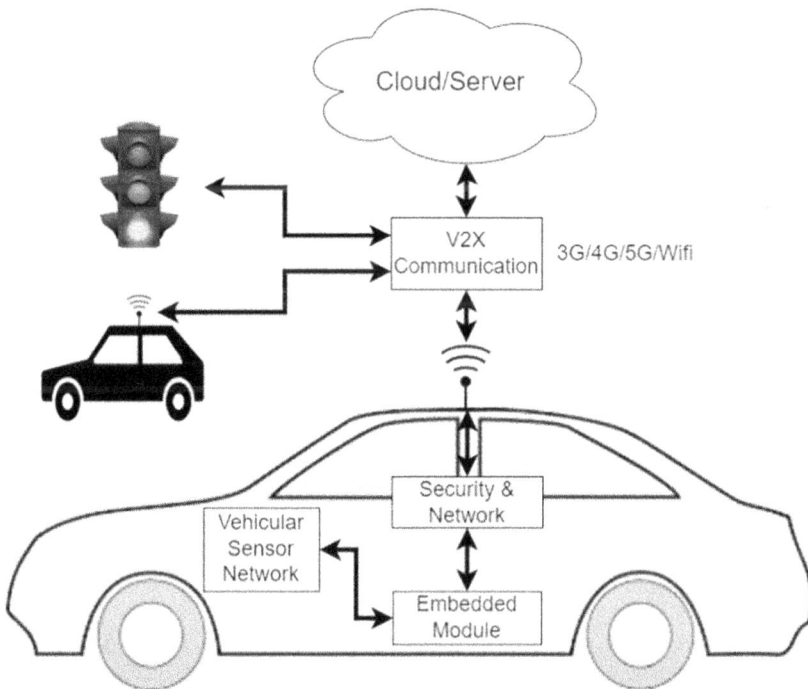

Figure 12.1 Connected vehicle as IOT AutoNode.

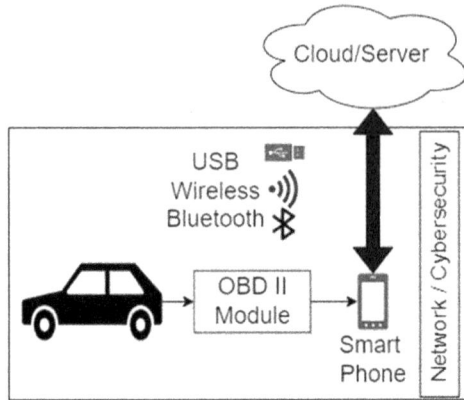

Figure 12.2 Non-connected vehicle conversion into an IOT AutoNode.

an AutoNode as represented in Figure 12.2. This utilizes low-cost hardware resources, principally comprising a data collection module and a connectivity interface [21]. In this case, the OBD-II module is used for primary vehicle data collection, whereas a smartphone enables access to the OBD data through an application in addition to providing inertial measurement unit (IMU) data and visual data through in-built cameras. Data acquired from the OBD module can be transferred to smartphones through wireless, USB or Bluetooth protocols. The simple low-cost implementation facilitates ease in access to hardware, improving usability across various demographics.

Vehicle parameters, combined with data acquired from smartphones, provide a wide range of features that can be used for vehicle and driving behavior analysis. The OBD module outputs data in two main forms: real-time vehicle data and diagnostics data.

Real-time vehicle data include parameters such as average and instantaneous speed, RPM, airflow rate, coolant temperature, accelerator, and brake pedal positions etc. By contrast, diagnostics data provide information about vehicle parameters at the time of troublesome events, including diagnostic trouble codes (DTCs). Similarly, several forms of data can be acquired through the smartphone interface due to its internal IMU comprising of sensors such as accelerometer, gyroscope, and magnetometer. As smartphones in the modern age are equipped with good resolution cameras, visual data during driving tasks act as a notable addition (Figure 12.3).

The OBD, sensor and visual data obtained from the vehicle can be pre-processed locally through on-board computing components and passed on to the cloud for further computation. As the data set contains several data types, the processing needs to be performed either sequentially or in parallel using triggers between data within the algorithm. The data obtained can be broadly classified into two types: vehicle-sensor data and visual data. ML and DL algorithms play an important role in data processing as the relationship between

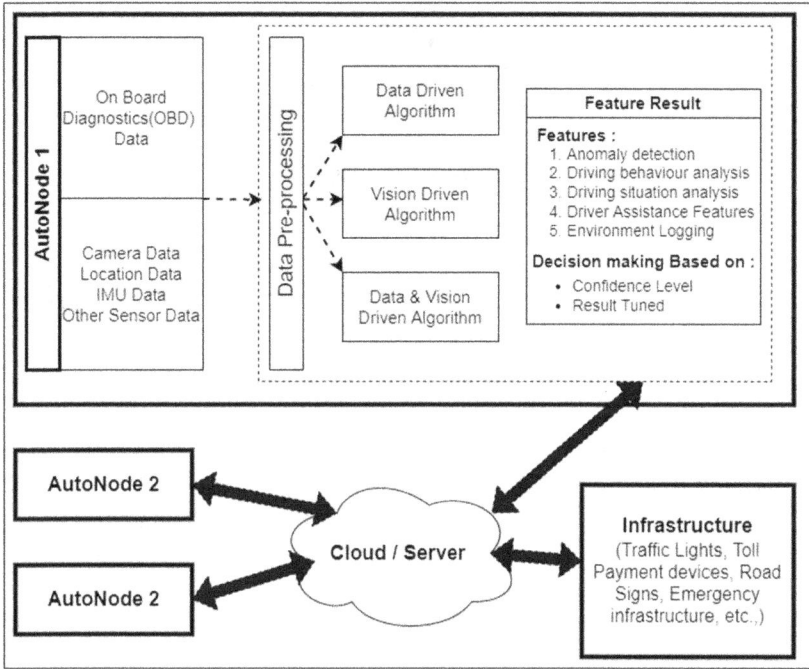

Figure 12.3 Flowchart of overall methodology of the framework [22].

these data types needs to be established. Convolutional Neural Network (CNN)-based DL architectures can be used to process visual data. Similarly, classification-based ML algorithms can be used for the post-processing of vehicle-sensor data [22].

The framework proposed in this chapter suggests crowdsourcing vehicular and visual data. However, supervised algorithms require massive amounts of data to achieve efficient results. In such cases, methods of transfer learning and reinforcement learning are implemented. Pre-trained CNN-based DL algorithms, such as YOLO (You Only Look Once) and VGG (Visual Geometry Group), are popular choices where image data is concerned [23, 24]. The major computation of visual and vehicle data takes place in the cloud, which is also responsible for the crowdsourcing and data redistribution to various entities establishing V2X communication.

The inclusion of external sensors and computational equipment in non-connected vehicles evolves the architecture to a connected vehicle architecture that is capable of exchanging data with the cloud. The cloud, in turn, is capable of sharing data with viable entities, including infrastructure and other vehicles. This implies that the difference between the architectures in terms of physical implementation lies only in the hardware setup. Similarly, the cloud plays a critical role in the algorithm component of the non-connected AutoNodes architecture where vehicles would not be able to

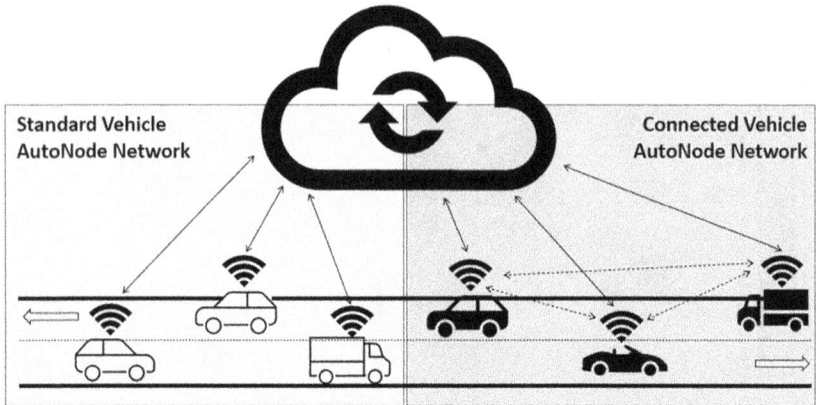

Figure 12.4 AutoNode network of standard and connected vehicle configurations.

communicate directly with each other in the absence of cloud communication, as shown in Figure 12.4. For in-vehicle communication networks, ECUs are the major entry points for cybercriminals whereas, in applications of V2X, connectivity devices such as transceivers and network routers are the major vulnerabilities. In this case, vehicular cybersecurity and standards gain a considerable amount of importance as data communication between vehicles and the cloud need to be protected.

12.3 APPLICATIONS AND FEATURES OF THE FRAMEWORK

AutoNodes that are implemented through CVs or non-connected vehicles provide an initiative-taking solution for several supporting applications, including road safety, traffic and fleet management, ADAS and vehicle maintenance. The exponential increase of urbanization demands road safety and traffic congestion solutions as they have serious consequences toward the safety of humans as well as increased infrastructure costs. On similar grounds, vehicle diagnostics and prognosis helps in understanding vehicle health and aid in environmental wellbeing. The framework proposed in this chapter benefits multiple stakeholders as the data is predominantly crowd-sourced. Among the stakeholders are the following:

- Vehicle manufacturers: Major automotive companies in the industry manufacture vehicles based on geographical requirements. Through AutoNodes, a comparison of performance of vehicles under different conditions can provide insightful data in understanding the requirement of vehicular features in a geographical location.
- Government: The government can improve infrastructure management and regular maintenance through the data acquired from AutoNodes.

- Incident analysis: Dangerous situation and accident analysis can be performed through data from AutoNodes. Traffic-related updates and alerts can also form an important component of incident analysis.
- Traffic violations: The traffic department largely benefits from the implementation of AutoNodes as violations can be recorded and reported in real time supported by data, that would in turn automate billing and fining systems.
- Vehicle rental and insurance agencies: Driving scores can be formulated by using the driving behavior data obtained from AutoNodes.
- Occupant safety and allied services: In-vehicle visual data can be used to analyse occupant activity and a response to various driving styles. This data can be used to enhance safety features for infants and the elderly in vehicular environments.

Connected vehicles which are equipped with required sensors might help achieve the above-discussed applications to a high level. But non-connected vehicles, which are the major occupants of the market, cannot benefit from the CV network. Using the current framework, our previous research was performed on four different non-connected vehicles with different drivers and locations. The inferences that are made from the framework [21, 22] have been classified into five main categories, as shown in Figure 12.3:

- Anomaly detection
- Driving behavior analysis
- Driving situation analysis
- Driver assistance features
- Environment logging

The feature of anomaly detection uses the data from IMU and the camera sensors of the smartphone for the detection of potholes, as shown in Figure 12.5(a), whereas the OBD interface data of the car are used for the analysis of sensor data of the vehicle to extract the speed of the car, the accelerator pedal position, etc. Accelerometer data perception of pothole detection in the current approach can be observed in Figure 12.5(a). The data undergoes filtering and thresholding methods to render negative peaks, where the vehicle encounters the pothole. A trigger-based system is developed based on data-driven and image-driven results that are used for validation. After validation, the location of potholes has been extracted in the form of longitude and latitude and plotted on a map for representation, as marked in Figure 12.6. Real-time vehicular data obtained from the OBD module, such as vehicle speed and accelerator pedal position, shown in Figure 12.5(b) and (c), are utilized to understand ride quality.

The driving behavior feature has a number of sub-features that can be further improved by considering precautionary suggestions based on vehicle parameters on regular routes. Parameters such as catalytic converter, overall fuel consumption and vehicle speed have been used for this purpose.

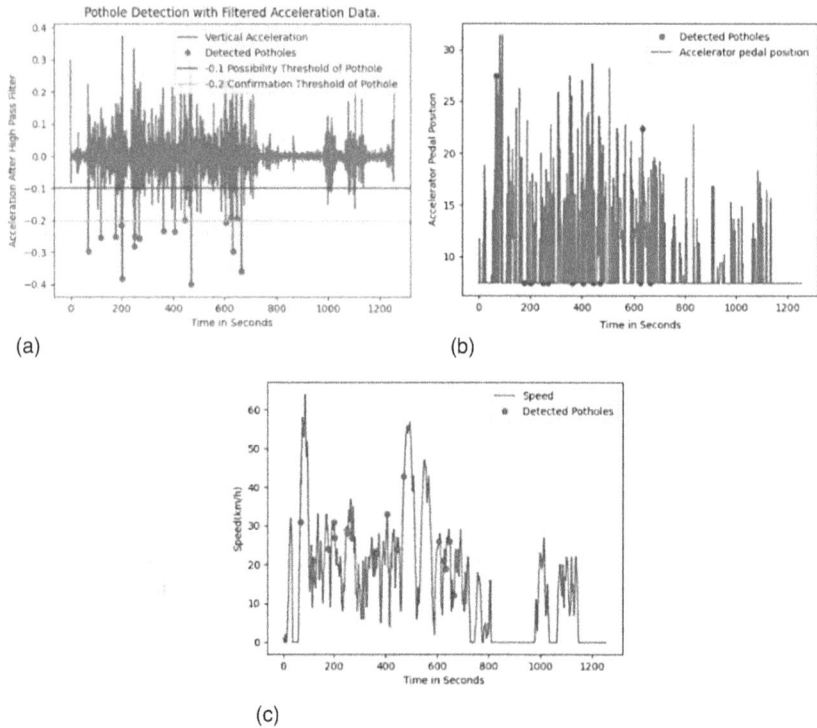

Figure 12.5 (a) Pothole validation through accelerometer data (b) Accelerator pedal position and speed through potholes (c) Values of speed of vehicle with respect to pothole presence on the road [21, 22].

The performance of the catalytic converter depends on the mean O_2 sensor values from the vehicle obtained from the OBD module. On similar grounds, average fuel consumption during the drive is established using the "L/100km" method. One important sub-feature of driving behavior analysis is the recognition of over-speeding instances during the drive. The parameters used to detect these instances are torque, vehicle speed and engine RPM. The parameters compass bearing, along with vehicle speed values, help in detecting sharp turns which forms another sub-feature (Figure 12.6).

Driving situation analysis in this context focuses on traffic detection and analysis based on OBD and visual data. The data-driven approach indicates a reduction in speed whereas the image data processed using the YOLO-V3 architecture is used to detect classes of objects on the road. The pre-trained model detects more than 80 classes of objects, including stop signs, cars, motorcycles, pedestrians, buses, and animals. The feature extractor in the YOLO-V3 architecture is DarkNet-53, which is 1.5 times faster when compared to previous versions. Situational analysis is performed based on these parameters, as shown in Figure 12.7(b).

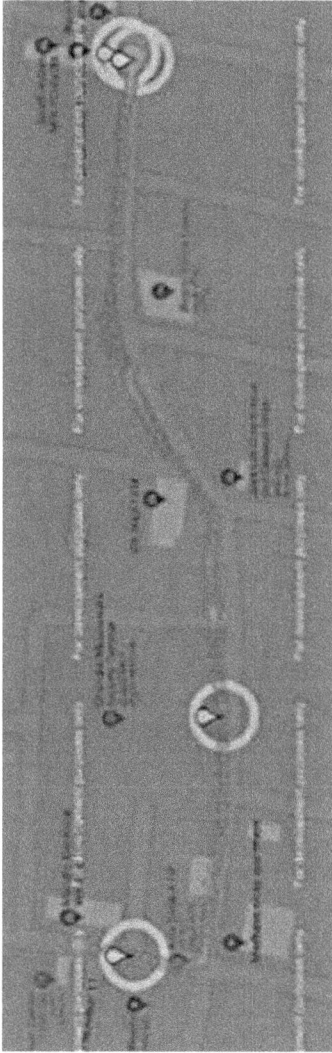

Figure 12.6 Geotagged locations of potholes detected [21].

(a)

(b)

Figure 12.7 (a) Pothole detected through smartphone camera sensor (b) Result of situational driving analysis [21, 22].

12.4 CONCLUSION

The chapter talks about the overview of automotive electronic systems and the evolution of OBD to establish the scale of data generated by automobiles. The complexity of systems has increased multifold and the possibilities of applications at scale using the huge amount of data provided by vehicles has been established. Advanced features such as ADAS, diagnostics, applications using image processing, etc., are discussed in detail. The proposed framework conclusively, with the help of the results presented, establishes that non-connected vehicles can have multiple advanced features of con-nected vehicles and enable any AutoNode to have ADAS along with the aspects of standards and security. The discussed methodology emphasizes on the applications of taking previous research by authors on road anomaly detection, behavioral analysis etc. Further extension to the current meth-odology with the implementation of AI, ML and DL algorithms trained with the help of collected data, can bring in a lot of additional inferences. AutoNodes, along with V2X infrastructure established on the large scale, will have a very substantial impact on the advanced features of an automo-bile in achieving safe, eco-friendly, and efficient drive cycles.

REFERENCES

[1] R. Menaka and K. Ashath, "A study on role of automobile industry in India and its customers satisfaction," *Shanlax International Journal of Management*, vol. 2, no. 4, pp. 2321–46433, 2015.

[2] A. Oktav, "New trends and recent developments in automotive engineering chapter 331 new trends and recent develelopments in automotive engineering Akın OKTAV." [Online]. Available: https://www.researchgate.net/publication/321621798

[3] "J3016_202104: Taxonomy and Definitions for Terms Related to Driving Automation Systems for On-Road Motor Vehicles – SAE International." https://www.sae.org/standards/content/j3016_202104/ (accessed Jul. 18, 2022).

[4] A. Ziebinski, R. Cupek, D. Grzechca, and L. Chruszczyk, "Review of advanced driver assistance systems (ADAS)," *AIP Conference Proceedings*, vol. 1906, 2017, doi: 10.1063/1.5012394.

[5] B. J. Shinde, S. S. Kore, and S. S. Thipse, "Comparative study of on board diag-nostics systems-EOBD, OBD-I, OBD-II, IOBD-I and IOBD-II," *International Research Journal of Engineering and Technology*, 2016, Accessed: Jul. 18, 2022. [Online]. Available: www.irjet.net

[6] "Ministry of Shipping, Road Transport and Highways G.S.R 84(E)," Feb. 9, 2009. https://morth.nic.in/sites/default/files/notifications_document/Notification_No__S_O_84E_dated_09-02-2009_Part-1_0.pdf (accessed Jul. 18, 2022).

[7] "India Bharat Stage VI Emission Standards," *ICCT POLICY UPDATES*, Apr. 2016. https://theicct.org/sites/default/files/publications/India%20BS%20VI%20Policy%20Update%20vF.pdf (accessed Jul. 18, 2022).

[8] Y. Ko and K. Hosoi, "Measurements of power losses in automobile drive train," *SAE Technical Papers*, 1984, doi: 10.4271/840054.

[9] C. M. C. G. Fernandes, P. M. T. Marques, R. C. Martins, and J. H. O. Seabra, "Gearbox power loss. Part II: Friction losses in gears," *Tribology International*, vol. 88, pp. 309–316, Apr. 2015, doi: 10.1016/J.TRIBOINT.2014.12.004.

[10] D. Singh and M. Singh, "Internet of vehicles for smart and safe driving," *2015 International Conference on Connected Vehicles and Expo, ICCVE 2015 – Proceedings*, pp. 328–329, 2016, doi: 10.1109/ICCVE.2015.93.

[11] K. Jadaan, S. Zeater, and Y. Abukhalil, "Connected vehicles: An innovative transport technology," *Procedia Engineering*, vol. 187, pp. 641–648, 2017, doi: 10.1016/J.PROENG.2017.04.425.

[12] M. Pinelis, "Automotive sensors and electronics: trends and developments in 2013," *Sensors Electronics Expo*, vol. 2013, 2013.

[13] N. Lu, N. Cheng, N. Zhang, X. Shen, and J. W. Mark, "Connected vehicles: Solutions and challenges," *IEEE Internet of Things Journal*, vol. 1, no. 4, pp. 289–299, Aug. 2014, doi: 10.1109/JIOT.2014.2327587.

[14] Y. Yang, D. Fei, and S. Dang, "Inter-vehicle cooperation channel estimation for IEEE 802.11p V2I communications," *Journal of Communications and Networks*, vol. 19, no. 3, pp. 227–238, 2017, doi: 10.1109/JCN.2017.000040.

[15] P. Sharma, D. Austin, and H. Liu, "Attacks on Machine Learning: Adversarial Examples in Connected and Autonomous Vehicles," *2019 IEEE International Symposium on Technologies for Homeland Security, HST 2019*, 2019, doi: 10.1109/HST47167.2019.9032989.

[16] M. Raya and J. P. Hubaux, "The Security of Vehicular ad Hoc Networks," *SASN'05 - Proceedings of the 2005 ACM Workshop on Security of Ad Hoc and Sensor Networks*, vol. 2005, pp. 11–21, 2005, doi: 10.1145/1102219.1102223.

[17] P. Gliwa, "Safety and ISO 26262," *Embedded Software Timing*, pp. 289–295, 2021, doi: 10.1007/978-3-030-64144-3_11.

[18] "ISO - ISO/PAS 21448:2019 – Road vehicles – Safety of the intended functionality." https://www.iso.org/standard/70939.html (accessed Jul. 18, 2022).

[19] M. Tschersich, "How to prepare automotive for future challenges iso 21434 – a standard for cybersecurity engineering," *VDI Berichte*, vol. 2017, no. 2310, pp. 29–30, 2017, doi: 10.51202/9783181023105-29.

[20] "India Connected Car Market Size, Share, Forecast Report – 2025." https://www.marketsandmarkets.com/Market-Reports/india-connected-car-market-15076 5405.html (accessed Jul. 18, 2022).

[21] K. S. Ashwini, G. Bhagwat, T. Sharma, and P. S. Pagala, "Trigger-Based Pothole Detection Using Smartphone and OBD-II," *Proceedings of CONECCT 2020 - 6th IEEE International Conference on Electronics, Computing and Communication Technologies*, 2020, doi: 10.1109/CONECCT50063.2020.9198602.

[22] S. Sabde, P. S. Pagala, K. Mudaliar, and K. S. Ashwini, "Scalable Machine Learning and Analytics of the Vehicle Data to derive Vehicle Health and Driving Characteristics," *Proceedings of IEEE International Conference on Disruptive Technologies for Multi-Disciplinary Research and Applications, CENTCON 2021*, pp. 146–151, 2021, doi: 10.1109/CENTCON52345.2021.9687913.

[23] W. Lan, J. Dang, Y. Wang, and S. Wang, "Pedestrian Detection Based on YOLO Network Model," in *IEEE International Conference on Mechatronics and Automation*, 2018.

[24] S. Arjapure and D. R. Kalbande, "Deep Learning Model for Pothole Detection and Area Computation," in *Proceedings – International Conference on Communication, Information and Computing Technology, ICCICT 2021*, 2021. doi: 10.1109/ICCICT50803.2021.9510073.

Chapter 13

An array of Fibonacci series-based wide-band printed antennas for IoT/5G applications

Kalyan Sundar Kola and Anirban Chatterjee
National Institute of Technology, Ponda, Goa

CONTENTS

13.1 INTRODUCTION

As internet connection spreads among many devices, the Internet of Things (IoT) has become an increasingly popular subject for discussion. Ericsson claims that by 2025, more than 5 billion gadgets will simultaneously be linked to the internet. Over the same period, 5G is becoming increasingly popular as a replacement for LTE (4G), which suffers from low data rates, limited bandwidth, and poor quality of service (QoS) in metropolitan areas. 5G, by contrast, delivers fast data rates, substantial capacity, and excellent spectrum efficiency. Outdoor, urban, and interior systems will all be needed for the new 5G technology. Accordingly, IoT and 5G technologies are in high demand because of these benefits, whose architecture is depicted in Figure 13.1(a) and (b), respectively. For 5G technology, low-cost and light-weight printed antennas [1, 2] are in high demand in this communication age for the connection of IoT devices. For this reason, a wide-band microstrip

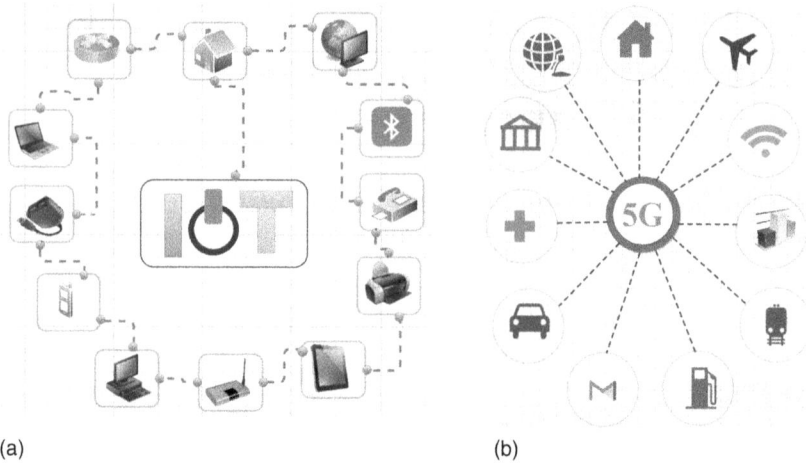

(a)   (b)

Figure 13.1 Architectural representation of (a) IoT-based applications (b) 5G communications.

patch antenna is an ideal choice for such uses. The printed antenna and its array have been studied in a number of publications.

In literature [3], printed antennas were studied by Carver et al. in both theoretical and practical terms. Furthermore, Werner et al. [4] study the progress of using fractal geometry in antenna design. The authors of that study developed antenna arrays using fractal geometries [4]. In another study, Al-Sehemi et al. [5] have suggested and fully analyzed a broadband waterproof antenna based on IoT applications and Ashyap et al. [6] developed a fabric material laminated C-shaped printed antenna for medical IoT applications. Cowsigan et al. [7] have presented a SIW cavity-backed printed antenna for IoT applications and an ESP8266 antenna module for IoT applications is provided by Roges et al. [8]. As described by Sharma et al.[9], future wireless applications may well benefit from a six-element MIMO antenna. In the IoT context, Mushtaq et al. [10] proposed an array of T-shaped slotted printed antennas. According to Wang et al. [11], a circularly polarized printed meta-material antenna for 5G indoor applications has been developed. An IoT application-oriented printed antenna that is 5G-enabled has been built and thoroughly tested by De et al. [12] And Singh et al. [13] introduced a dual-band microstrip patch antenna for 5G applications based on a SIW antenna structure. In addition, many studies have suggested using an antenna array to enhance parametric outcomes. Wang et al. [14] introduced a planar array of U-slotted patch antennas. Clover-leaf-shaped geometries inspired by nature have been published by Kola et al. [15]. The author has also developed and analyzed a variety of microstrip patch antenna arrays, such as Christmas-tree, hybrid fractal, and tulip flower geometries [16–19]. Drawing on this background, this chapter reports an IoT/5G applications-based 2-element linear printed array of monopole antennas.

Among the new features of the proposed antennas are the following:

- The sole antenna is derived from the Fibonacci series of successive seventh terms.
- The sole antenna is capable of offering wide impedance bandwidth of 2.3 GHz.
- It also offers below −35dB of x-pol. suppression along the main lobe of direction.
- It also offers quite good peak-gain at both of the resonating frequencies.
- The array's feed network has been designed using the Wilkinson power splitter [20] to achieve low loss, high isolation, and improved bandwidth responsiveness.
- In addition, the array has a broad impedance bandwidth, a high gain, and a shallow x-pol. level at the ideal radiation point.

In addition to the advantages mentioned above, the single antenna and the array have aperture [24, 25] and radiation efficiencies [26–30] of over 63% and 92%, respectively. At appropriate resonating frequencies, both antennas give more than 30 dB/m correction factors [31] and so have modest electromagnetic interference effects. As a result of their specifications, both antennas are appropriate for use in IoT and 5G networks.

13.2 PROPOSED SINGLE ANTENNA

Seven terms of the Fibonacci sequence were considered in the development of the single antenna. Detailed geometry creation, antenna construction, perimeter and area calculations, and time-domain analysis have all been completed here.

13.2.1 Formation of the proposed geometry

The step-wise formation of the proposed 'Fibonacci series'-based geometry is described in Figure 13.2. There are seven successive Fibonacci squares in the one quadrant construction, and the side length is calculated [21] as follows.

$$k_{S+1} = \varphi \times k_S, 1 \le s \le 6 \tag{13.1}$$

where, φ is the 'golden ratio' (≈ 1.67) and $k_1 = 1$ mm.

Initially, a square geometry **A** is considered whose side length is kept as k_1, which is shown in Figure 13.2(a). In step-1, the square **A** is scaled with φ

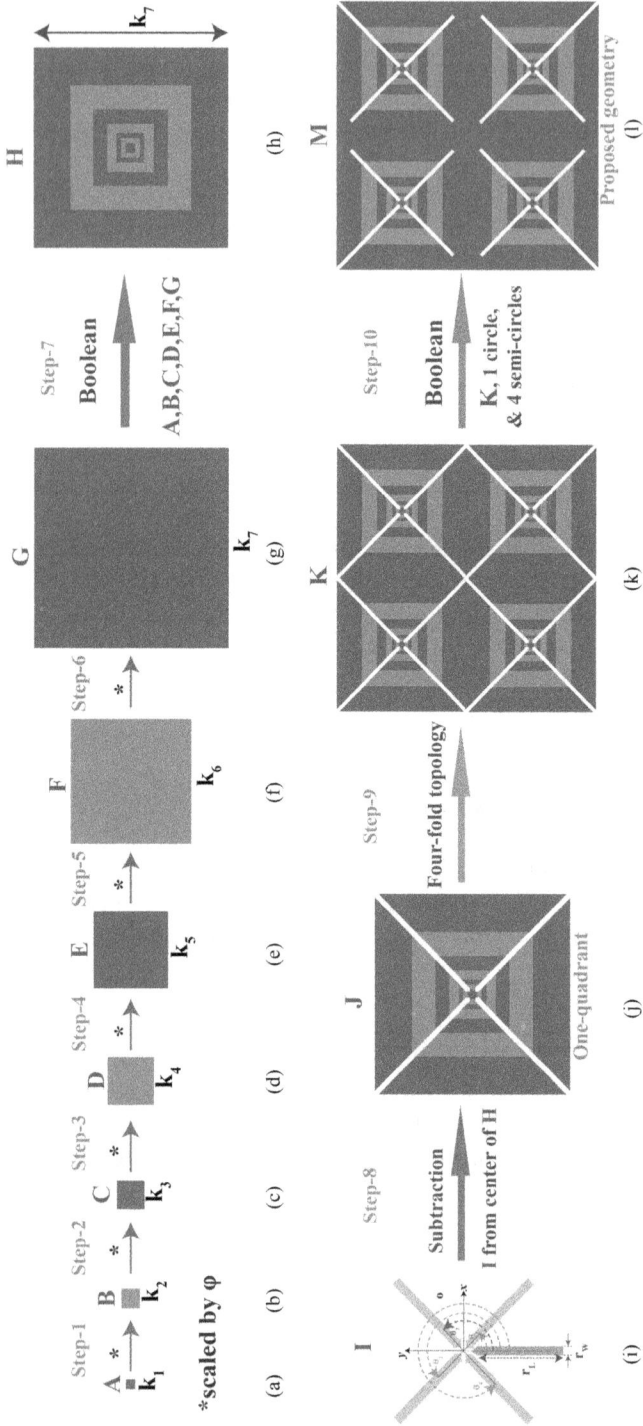

Figure 13.2 Step-wise formation of proposed Fibonacci series-based geometry.

and this obtains the scaled geometry **B**, whose side length becomes k_2 which is presented in Figure 13.2(b). Further, the geometry **B** is scaled by φ and obtains the resultant geometry **C** after the completion of step-2, whose side length is k_3 as shown in Figure 13.2(c). Similarly, the geometries **D, E, F** and **G** have been achieved following the completion of steps-3,4,5,6, and those are represented in Figure 13.2(d)–(g), respectively. The side length of the resultant geometries, achieved from steps-3,4,5,6 becomes k_4, k_5, k_6, k_7, respectively. In step-7, all the resultant geometries obtained from the preceding steps are Boolean and they have achieved the Fibonacci square-based geometry **H** whose overall side length becomes k_7. This is shown in Figure 13.2(h). Further, a structure **I** is formed by the combination of four arrow shaped geometries where the arrows make an angle of θ_1, θ_2, θ_3, θ_4 with respect to the generating arrow, as shown in Figure 13.2(i). The length and width of the thinned arrow geometry are fixed as r_L and r_W, respectively, as indicated in Figure 13.2(i). In order for the structure to resonate at the desired frequency, the geometry **I** was cut out of each corner of **H**. After step-8, we got the one-quadrant Fibonacci square-based slotted structure **J**, which is shown in Figure 13.2(j). To produce a symmetric radiation pattern, the four values of **J** are positioned in a four-quadrant of the xy-plane and then overlaid them to form the 2D-symmetric structure **K**, as shown in Figure 13.2(k). To make sure that all of the parts of the geometry **K** are connected, a circle and four semicircles are Boolean with the **K**. This is how the final proposed geometry **M**, shown in Figure 13.2(l), was made. The numeric values of the used notations are enlisted in Table 13.1.

13.2.2 Depiction of single antenna

To build the printed antenna, as shown in Figure 13.3, the final recommended geometries of the step-wise construction are laminated on the top layer of Rogger 5880 PCB material. Figure 13.3(a) depicts the antenna's frontal perspective. A microstrip line feed has been used to excite the antenna construction, and the optimal length of the feed is L_f, as indicated in Figure 13.3(a). A partial ground plane is used underneath the substrate layer of the antenna in order to create a broad impedance bandwidth. The partial ground plane has also been etched with a rectangular slot ($S_L \times S_W$), which runs parallel to the feed line and ensures correct impedance matching

Table 13.1 Dimension details of the single antenna (unit: mm)

Exponent	Esteem	Exponent	Esteem	Exponent	Esteem	Exponent	Esteem
k_1	1	K_2	2	K_3	3	K_4	5
K_5	8	K_6	13	K_7	21	r_L	11
r_W	0.5	r	1	W	42	L	42
S_L	5	S_W	2.29	L_f	20.9	L_g	19.65

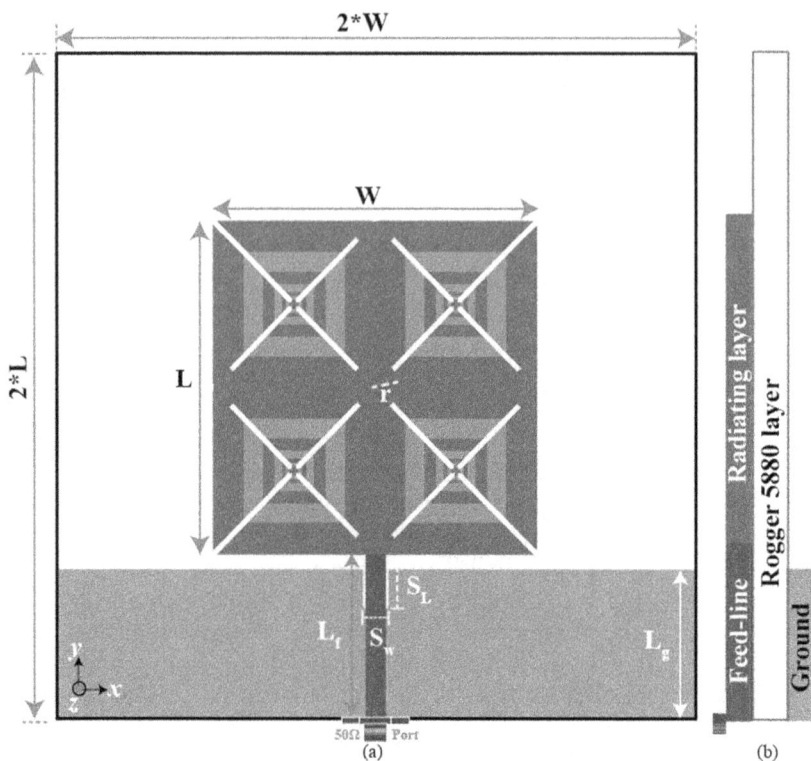

Figure 13.3 Proposed antenna (a) front-view (b) side-view.

at the required frequency range. Figure 13.3(a) shows that the radius of the connecting circle is set at **r**. The vertical view of the antenna is reflected in Figure 13.3(b). It is evident from the vertical perspective of the antenna that the suggested antenna is a single-layered construction. The proposed structure is simulated in the SONNET EM simulator [22], and, further, the results are cross-verified by CST microwave studio [23]. Table 13.1 lists all of the dimensions' numerical values.

13.2.3 Contour length and surface area computation

The effective perimeter and the surface area of the suggested structure can be calculated as follows:

Contour length:
The contour length of the superimposed geometry **H** is k_7 mm. Hence, the perimeter of **H** is computed as:

$$P_H = 4 \times k_7 \text{ mm} \tag{13.2}$$

In order to increase the electrical length of **J**, four slotted based geometry (**I**) has been etched from geometry **H** where the contour length of **I** is computed as follows:

$$P_I = 4 \times 2 \times (r_L + r_W) \, \text{mm} \tag{13.3}$$

Therefore, the effective contour length of the one-quadrant geometry **J** becomes:

$$P_J = P_H + P_I = 4 \times \{k_7 + 2(r_L + r_W)\} \, \text{mm} \tag{13.4}$$

Similarly, the contour length of the proposed four-quadrant geometry **K** is computed as follows:

$$P_K = 4 \times P_J \, \text{mm} \tag{13.5}$$

Also, the geometry **K** adds one circle and four semicircles, which make the structure shorter from an electrical point of view. The following equation shows how to figure out the effective length of the final geometry **M**'s contours:

$$P_M = P_K - (6 \times \pi \times r) \, \text{mm} \tag{13.6}$$

After putting all the parameter values in equations (13.2)–(13.6) and the computed effective contour length of the final proposed geometry becomes **685.15 mm.**

Surface area:
The radiating surface area of the suggested superimposed structure **H** is derived as follows:

$$A_H = K_7^2 \, \text{mm}^2 \tag{13.7}$$

After subtracting the structure **I**, the overall area of the one-quadrant structure **J** is reduced and it becomes

$$A_J = K_7^2 - (4 \times r_L \times r_W) \, \text{mm}^2 \tag{13.8}$$

Now the surface area of the four-folded geometry **K** is as follows:

$$A_K = 4 \times A_J^2 \, \text{mm}^2 \tag{13.9}$$

Further, one circle and four numbers of semi-circles are added with the geometry **K**, which improve the surface area of the structure. Therefore, the overall surface area of the final geometry **M** is computed as follows:

$$A_M = A_K + (3 \times \pi \times r^2) \, \text{mm}^2 \tag{13.10}$$

After putting all the parameter values in equations (13.7)–(13.10) and the computed surface area or the radiating of the final proposed geometry becomes **16.85 cm²**.

13.2.4 Time domain analysis

Time domain characteristics, such as input–output signal quality, scattering parameters, and group delay, are examined here. The identical antennas are placed in two different ways on the platform of CST microwave studio, as illustrated in Figures. 13.4(a) and (b), one horizontally and the other vertically. The inter-element spacing has been maintained at 0.3 meters for both arrangements. Figure 13.4(c) shows the desired output signal values when a sinusoidal signal is supplied to the antenna's port. Antenna isolation is shown in Figure 13.4(d) for two different configurations. As can be seen in Figure 13.4(d), excellent isolation of less than –22 dB was obtained from both arrangements throughout the entire input frequency range. As seen in Figure 13.4(e), the phase response of the antennas shows linear phase fluctuation, indicating a well-designed system. Figure 13.4(f) depicts the group delay characteristics visually. Group delay is below 2ns in both situations, which is desirable and widely acceptable for wideband applications, particularly for IoT and 5G communications.

13.3 PROPOSED ARRAY

This section explains the proposed power divider, the building of the array, and the benefits of the Wilkinson power divider network.

13.3.1 Construction of power divider network

A three-port power divider network [20] is presented in Figure 13.5. Basically, it's a Wilkinson power divider network where the source power is forwarded to the sink ports by maintaining an equal magnitude. It follows the transmission line theory principle. The power is driven from the source port using a transmission line whose characteristics impedance is Z_0 and transferred to the quarter-wavelength junction whose characteristics impedance is $\sqrt{2}Z_0$. Further, the power is equally delivered to the output ports through the Z_0 transmission lines. For isolation, the $2Z_0$ resistor is fixed in the quarter-wavelength junction, as shown in Figure 13.5.

The transmission line's [1] input impedance (Z_{in}) can be determined as follows:

$$Z_{in} = Z_0 \frac{Z_L + jZ_0 \tan \beta l}{Z_0 + jZ_L \tan \beta l} \tag{13.11}$$

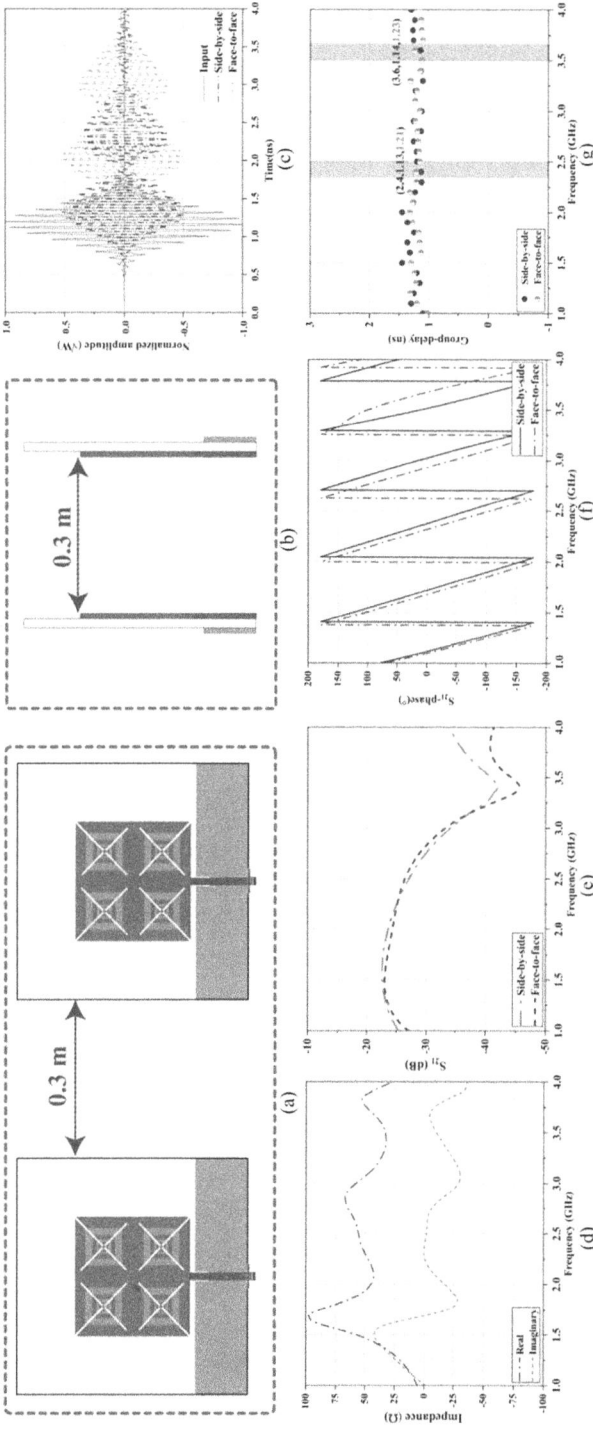

Figure 13.4 Time domain analysis of the single antenna: (a) side-by-side (b) face-to-face (c) input–output signal (d) impedance (e) S_{21} magnitude (f) S_{21} phase (g) group delay.

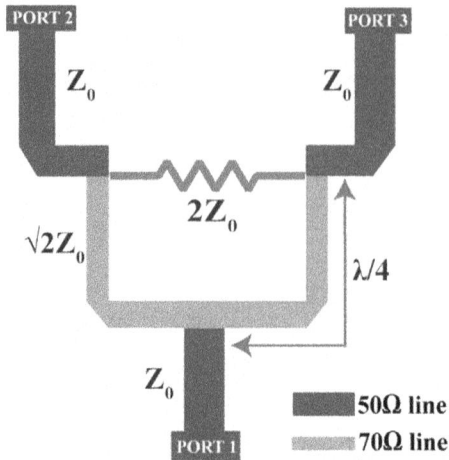

Figure 13.5 A 3-way power divider network.

where Z_L and Z_0 are indicated as load and characteristics impedance of the antenna and l is considered as the length of the transmission line, respectively.

13.3.2 Development of array

A two-element array, shown in Figure 13.6, has been suggested to enhance parametric results. It's a linear array with a 0.5 (**R**) separation between the elements. A shared partial ground plane is implemented to accomplish the broad impedance bandwidth necessary for IoT and 5G applications. As part of the array's feed network, the Wilkinson power divider is used. 50Ω–70Ω–50Ω transmission lines make up the network, which sends out an equal amount of power to all sink ports of the antennas. The 50Ω transmission line's length and width are set as $\lambda/2$ and 2.29 mm, respectively. The 70Ω transmission line's length and width are indeed an integer multiple of $\lambda/4$ and 1.31 mm, respectively. As seen in Figure 13.6, a quarter-wavelength transmission line's comprehensive description is reflected. Data about the transmission lines' dimension notations are presented in Table 13.2.

13.3.3 Benefits of the Wilkinson power divider

Two types of power divider networks may be utilized to build the array geometry: the conventional and the Wilkinson power divider networks. Using the Wilkinson power divider network, we may reduce the overall size of the array while still achieving the goals mentioned above of low loss [1] and high isolation [1]. Stable radiation patterns may also be achieved by

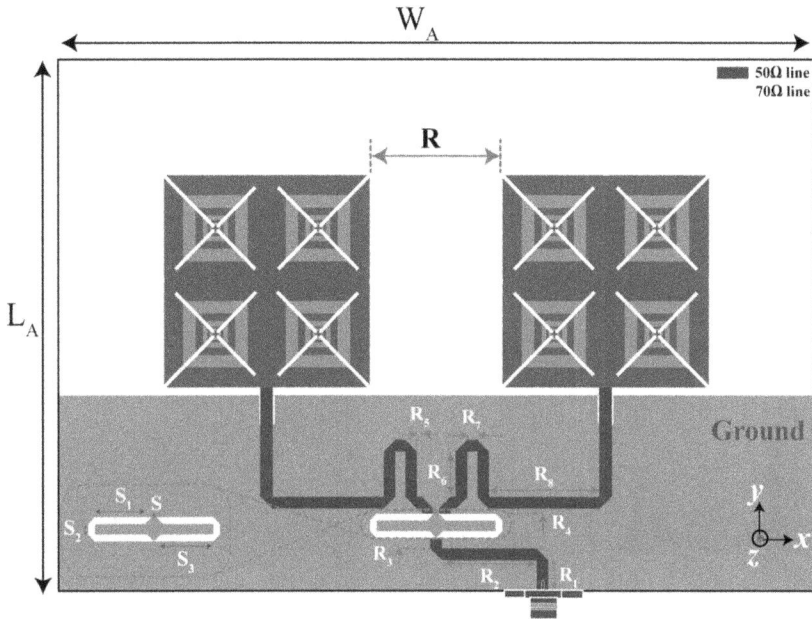

Figure 13.6 Proposed 2-element linear array geometry.

Table 13.2 Dimension details of the proposed array (unit: mm)

Exponent	Esteem	Exponent	Esteem	Exponent	Esteem	Exponent	Esteem
W_A	150.70	L_A	98.30	R_1	2	R_2	18.71
R_3	2.02	R_4	1	R_5	1	R_6	8.60
R_7	2	R_8	21.93	S_1	9.94	S_2	2
S_3	10.44	S	1	R	25.74	--	--

using this method. Figure 13.7 shows the results of a performance comparison between arrays using standard and Wilkinson feeds [20]. Figures 13.7(a) and (b) show the return-loss characteristics of a standard feed and its related arrays. Furthermore, the Wilkinson power divider network and its related arrays' return loss are shown in Figures 13.7(c) and (d). A Wilkinson power divider network may be plainly shown to provide a superior return loss and impedance bandwidth than an array that uses the standard feed. Table 13.3 displays the parametric results side by side in a tabular manner. Table 13.3 clearly shows that the Wilkinson power divider-based suggested array has superior parameters with regard to gain, radiation efficiency, and other metrics.

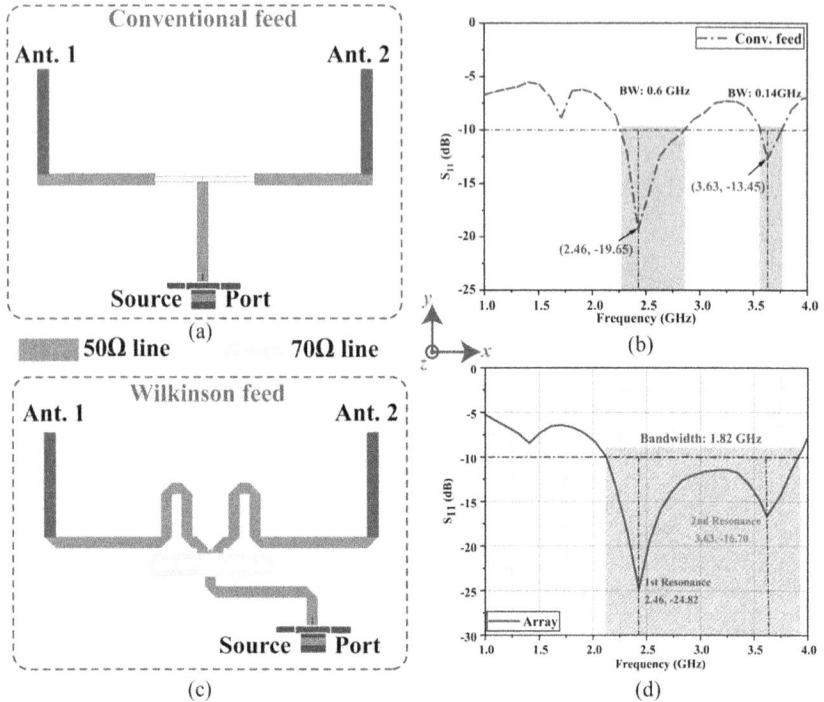

Figure 13.7 Comparative performance of array geometries (a) conventional feed, (b) resultant S_{11} of (a), (c) Wilkinson power feed (d) resultant S_{11} of (c).

Table 13.3 Comparative performance of the arrays

Feed-type	Size [mm²]	f_r [GHz]	S_{11} [dB]	BW [GHz]	Gain [dB]	η [%]
Conventional	150.70 × 98.30	2.46, 3.63	−19.65, −13.45	0.6, 0.14	4.22, 4.64	88, 87
Wilkinson	150.70 × 102.30	2.46, 3.63	−24.82, −16.70	1.82	7.17, 7.82	97, 93

13.4 RESULTS AND DISCUSSION

The proposed printed antenna and its array, based on the Fibonacci series and geared toward IoT/5G applications, have been modeled with the help of the SONNET EM simulator [22] and cross-verified by the CST microwave studio [23].

   The partial ground plane-based single antenna has resonated in two different frequencies, i.e., 2.40 and 3.60 GHz. It offers a satisfactory return loss of 32.56 and 19.12 dB, respectively, as depicted in Figure 13.8(a).

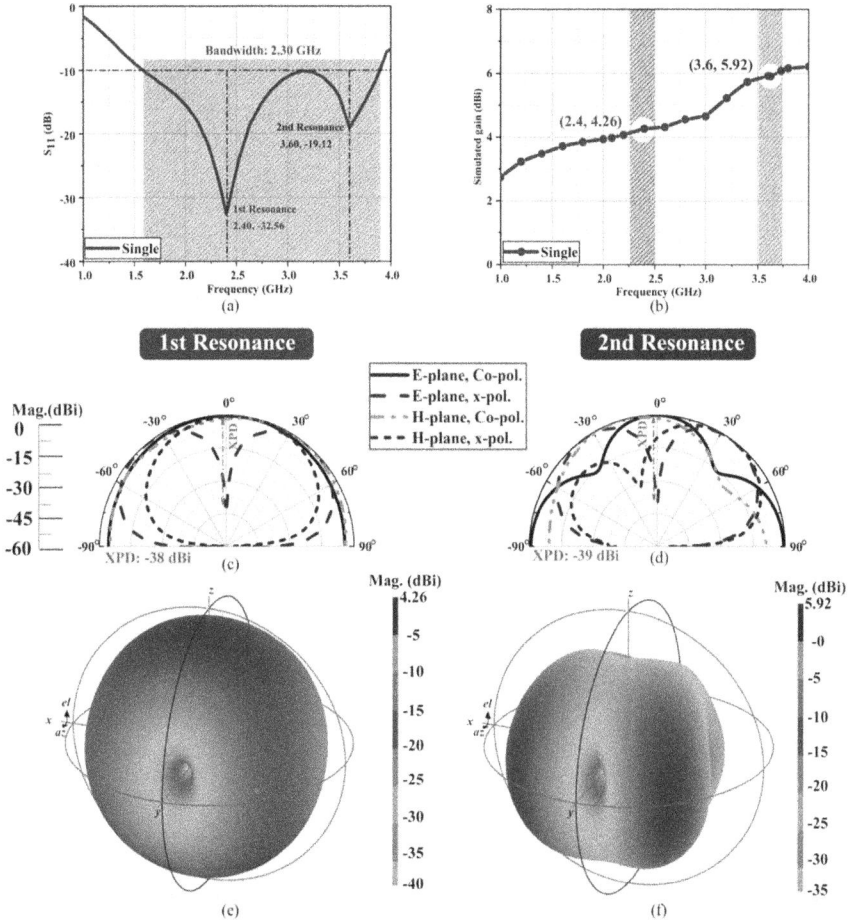

Figure 13.8 Simulated results of single antenna: (a) S_{11}, (b) gain (c) E- and H-field patterns for 1st resonance (d) E- and H-field patterns for 2nd resonance (e) 3D-radiation pattern of 1st resonance (f) (e) 3D-radiation pattern of 2nd resonance.

It also covers a wide-impedance bandwidth of 2.30GHz (1.55–3.85 GHz). The single antenna can satisfy the needs of IoT-based household appliances and 5G communications based on resonating frequencies. The sole antenna gives desired peak gain of 4.26 and 5.92 dBi at two resonances, as shown in Figure 13.8(b). The E- and H-field properties of the antenna for two reso-nating frequencies have been characterized in Figure 13.8(c) and (d), respec-tively. It can be clearly observed that the sole antenna gives <–35 dBi cross-pol. discrimination in both the frequencies, which indicates the excel-lent design. 3D radiation patterns of the antenna are presented in Figure 13.8(e) and (f), respectively, in which the right-hand scale indicates the max-imum magnitude achieved from the antenna. Additionally, it was discovered

Table 13.4 Parametric performances of the antenna and the array

Parameter(s)	Single antenna		Antenna array	
	1st Resonance	*2nd Resonance*	*1st Resonance*	*2nd Resonance*
f_r [GHz]	2.40	3.60	2.46	3.63
S_{11} [dB]	−32.56	−19.12	−24.82	−16.70
BW [GHz]	2.30		1.82	
Gain [dBi]	4.26	5.92	7.17	7.82
Directivity [dBi]	5.63	7.15	8.26	9.14
x-pol. [dBi]	−38	−39	−29	−32
Rad. eff. [%]	97	96	97	93
App. eff. [%]	74.90	64.77	75.27	63.68
CF [dB/m]	33.57	35.43	30.87	33.60

that the antenna produces a radiation pattern, which is symmetric in the required directions. Table 13.4 has an entry for every numeric value of the parameter. The single antenna gives 97% and 96% radiation efficiency for 2.40 and 3.60 GHz, respectively. It can be observed that the antenna is highly efficient for IoT and 5G applications.

An antenna's aperture efficiency [24, 25] is computed as follows:

$$\eta_{ap} = \frac{D}{D_{\max}} \tag{13.12}$$

$$D_{\max} = \frac{4 \times \Pi \times \text{Area}}{\lambda^2} \tag{13.13}$$

where D, Area are indicated as the directivity and overall surface area of the antenna, and λ is the free-space wavelength.

The overall surface area of the sole antenna is 83.8 × 83.8 mm² and the directivity is 5.63 and 7.15 dBi, respectively. As a result, the estimated aperture effectiveness of the antenna becomes 97% for a frequency of 2.40 GHz and 96% for a frequency of 3.60 GHz.

One of the most essential aspects of an antenna's electromagnetic impact is its electromagnetic compatibility (EMC). Using the surface equivalence technique, the antenna's correction factor (CF) [31] is determined, and the CF is then calculated as follows:

$$CF = \frac{\left|E^{\text{INC}}\right|}{\left|V_r\right|} = 20\log\left(\frac{9.73}{\lambda\sqrt{G_a}}\right) \tag{13.14}$$

where, $\left|E^{\text{INC}}\right| \rightarrow$ incident field strength, $\left|V_r\right| \rightarrow$ antenna terminal voltage, and $G_a \rightarrow$ antenna's gain. The single antenna gives the correction factor of 33.57

and 35.43 dB/m at two resonating frequencies, and these are widely acceptable for the desired applications.

With the help of the Fibonacci series-based single antenna, a two-element array has been proposed and developed to improve parametric outcomes. Figure 13.9(a) shows that the suggested array has a return loss of 24.82 dB and 16.70 dB, respectively, at 2.46 and 3.63 GHz. The array also has a large impedance bandwidth of 1.82 GHz, covering the entire IoT and 5G range. It can be seen in Figure 13.9(b) that the antenna offers 7.17 and 7.82 dBi gains, respectively, which are broadly accepted for the intended applications. In Figure 13.9(c) and (d), the E- and H-fields of the array are described. One noticeable feature of the array is that the x-pol. suppression is relatively

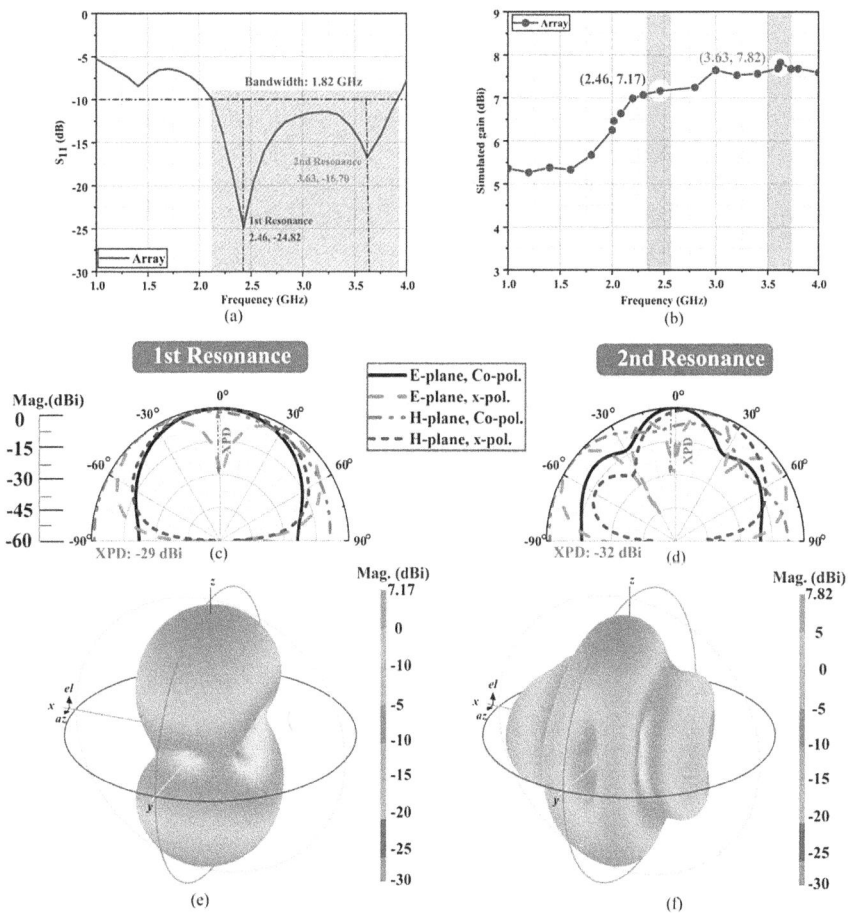

Figure 13.9 Simulated results of array: (a) S_{11}, (b) gain (c) E- and H-field patterns for 1st resonance (d) E- and H-field patterns for 2nd resonance (e) 3D-radiation pattern of 1st resonance (f) (e) 3D-radiation pattern of 2nd resonance.

shallow, with a value below –29 dBi that is desired and commonly accepted. Figure 13.9(e) and (f) show that the array provides a steady radiation pattern at the correct frequencies. The antenna array's simulated outcomes are enlisted in Table 13.4. The radiation efficiency of the array is shown to be 97% and 93%, respectively. The array's aperture is 150.7 × 98.3 mm² and its directivity is 8.26 and 9.14 dBi. For the two different frequencies of operation, the array's aperture efficiency was determined as 75.27% and 63.68%, respectively. The array's computed CF values are 30.87 and 33.6 dB/m, which are desired. Both antennas can support IoT-based home applications and 5G communications with acceptable performance parameters.

13.5 CONCLUSION

An array of a couple of printed radiators has been designed and its parametric behavior has been thoroughly analyzed. The single monopole radiator of the array is derived from the Fibonacci series based on seven square geometries. A four-fold architecture has been studied in order to generate symmetric radiation patterns at the necessary frequency. The resulting fourfolded shape was used to etch four numbers of thinned slotted structures, which provided the necessary resonant properties. The construction incorporates a partial ground plane with a rectangular slot in order to produce a broad impedance bandwidth. It has been analysed in the time domain to determine its signal quality, isolation, group delay, and other characteristics. The single antenna has a broad impedance bandwidth, strong gain, a shallow cross-pol. level, and great radiation efficiency. The two-element linear array was created to enhance parametric results. The Wilkinson power divider network is also used to construct its feed and achieve low loss, a wide impedance bandwidth, and intense isolation. The array's parametric findings are likewise promising. There is considerable potential for IoT-based home applications and 5G communications using these antennas.

ACKNOWLEDGEMENT

This chapter is an outcome of a project (File No. SB/S3/EECE/226/2016) under the SERB Extra Mural Research Funding, DST, GoI.

REFERENCES

[1] C.A. Balanis, *Antenna Theory: Analysis and Design*, Wiley, 1997.
[2] R.L. Haupt, *Antenna Arrays: A Computational Approach*, Wiley-IEEE Press, 2010.

[3] K.R. Carver, and J.W. Mink, "Microstrip Antenna Technology," *IEEE Transactions on Antennas and Propagation*, vol. AP-29(1), pp. 2–24, 1981.

[4] D.H. Werner and S. Ganguly, "An Overview of Fractal Antenna Engineering Research," *IEEE Antennas and Propagation Magazine*, vol. 45(1), pp. 38–57, 2003.

[5] G.A. Abdullah, A.A. Ahmed, T. D. Nikolay, T.A. Nikolay, and L. A. Gabriela, "Design of a Flexible Waterproof Antenna for Internet of Things Applications," *Journal of Electromagnetic Waves and Applications*, vol. 35(7), pp. 874–887, 2021.

[6] Y.I.A. Adel et al., "C-shaped Antenna Based Artificial Magnetic Conductor Structure for Wearable IoT Healthcare Devices," *Wireless Networks*, vol. 27, pp. 4967–4985, 2021.

[7] S.P. Cowsigan, D. Saraswady, "Substrate Integrated Waveguide Cavity Backed Antenna for IoT Applications," *Journal of Ambient Intelligence and Humanized Computing*, pp. 1–6, 2021. doi: 10.1007/s12652-020-02742-0.

[8] R. Roges, and P.K. Malik, "Planar and Printed Antennas for Internet of Things-enabled Environment: Opportunities and Challenges," *International Journal of Communication Systems*, vol. 34(15), pp. 1–32, 2021.

[9] D. Sharma, B.K. Kanaujia, and S. Kumar, "Compact Multi-standard Planar MIMO Antenna for IoT/WLAN/Sub-6 GHz/X-band Applications," *Wireless Networks*, vol. 27, pp. 2671–2689, 2021.

[10] A. Mushtaq, A. Rajawat, and S.H. Gupta, "Design of Antenna Array Based Beam Repositioning for IoT Applications," *Wireless Personal Communications*, vol. 122, pp. 3205–3225, 2022.

[11] Z. Wang, T. Liang, and Y. Dong, "Metamaterial-Based, Compact, Wide Beam-Width Circularly Polarized Antenna for 5G Indoor Application," *Microwave and Optical Technology Letters*, vol. 63(8), pp. 2171–2178, 2021.

[12] A. De, B. Roy, and A.K. Bhattacharjee, "Miniaturized Dual Band Consumer Transceiver Antenna for 5G-Enabled IoT-Based Home Applications," *International Journal of Communication Systems*, vol. 34(11), pp. 1–14, 2021.

[13] U. Singh, and R. Mishra, "A Dual-Band High-Gain Substrate Integrated Waveguide Slot Antenna for 5G Application," *Progress In Electromagnetics Research C*, vol. 119, pp. 191–200, 2022.

[14] H. Wang, X.B. Huang, and D.G. Fang, "A Single Layer Wideband U-Slot Microstrip Patch Antenna Array," *IEEE Antennas and Wireless Propagation Letters*, vol. 7, pp. 9–12, 2008.

[15] K.S. Kola and A. Chatterjee, "A Two-element Array with Clover-leaf Shaped Antennas for X-band Applications," *7th International Conference on Signal Processing and Integrated Networks (SPIN)*, Noida, India, pp. 349–354, 2020.

[16] K.S. Kola and A. Chatterjee, "A Linear Array of Christmas-tree Shaped Antennas for DBS Applications," *International Conference on Computer, Electrical & Communication Engineering (ICCECE)*, Kolkata, India, pp. 1–6, 2020.

[17] K. S. Kola and A. Chatterjee, "A Printed Array of High-gain Fractal Antennas for X-band Applications," *International Conference on Communication, Computing and Industry 4.0 (C2I4)*, Bangalore, India, pp. 1–6, 2020.

[18] K. S. Kola and A. Chatterjee, "A 1 x 2 Array of High-gain Radiators for Direct Broadcast Satellite (DBS) Services under Ku-band," *8th International Conference on Signal Processing and Integrated Networks (SPIN)*, pp. 297–302, 2021.

[19] K. S. Kola and A. Chatterjee, "An Array of Tulip-flower Shaped Printed Radiators for Direct Broadcast Satellite (DBS) Applications," *Advanced Communication Technologies and Signal Processing (ACTS)*, vol. 2021, pp. 1–6, 2021.

[20] E.J. Wilkinson, "An N-way Hybrid Power Divider," *IRE Transactions on Microwave Theory Techniques*, vol. 8(1), pp. 116–118, 1960.

[21] E.O. Kreyszig, *Advanced Engineering Mathematics*, New York, Wiley, 1983.

[22] High Frequency Electromagnetic Software SONNET (v14), North Syracuse, NY 13212, April, 2013.

[23] CST STUDIO SUITE, (v2018), CST AG, Germany, 2018.

[24] Z. Ma, and A.E. Vandenbosch, "Low-Cost Wideband Microstrip Arrays With High Aperture Efficiency," *IEEE Transactions on Antennas and Propagation*, vol. 60(6), pp. 3028–3034, 2012.

[25] A. Vosoogh, and P.S. Kildal, "Simple Formula for Aperture Efficiency Reduction Due to Grating Lobes in Planar Phased Arrays," *IEEE Transactions on Antennas and Propagation*, vol. 64(6), pp. 2263–2269, 2016.

[26] E.H. Newman, P. Bohley, and C.H. Walter, "Two Methods for the Measurement of Antenna Efficiency," *IEEE Transactions on Antennas and Propagation*, vol. AP-23(4), pp. 457–461, 1975.

[27] D.M. Pozar, and B. Kaufman, "Comparison of Three Methods for the Measurement of Printed Antenna Efficiency," *IEEE Transactions on Antennas and Propagation*, vol. 36(1), pp. 136–139, 1988.

[28] G.S. Smith, "An Analysis of the Wheeler Method for Measuring the Radiating Efficiency of Antennas," *IEEE Transactions on Antennas and Propagation*, vol. 25(4), pp. 552–556, 1997.

[29] R. Chair, K.M. Luk, and K.F. Lee, "Radiation Efficiency Analysis on Small Antenna by Wheeler Cap Method," *Microwave and Optical Technology Letters*, vol. 33(2), pp. 112–113, 2002.

[30] M.A. Moharram, and A.A. Kishk, "MIMO Antennas Efficiency Measurement Using Wheeler Caps," *IEEE Transactions on Antennas and Propagation*, vol. 64(3), pp. 1115–1120, 2016.

[31] C.R. Paul, *Introduction to Electromagnetic Compatibility*, 2nd edition, Wiley, 2006.

Chapter 14

Deep learning IoT platform for dental disease detection

Sindhu P. Menon

School of Computing and Information Technology, REVA University, Bangalore, India

Pramod Kumar Naik and Baskar Venugopalan

Department of Computer Science and Engineering, Dayananda Sagar University, Bangalore, India

CONTENTS

14.1 INTRODUCTION

Dental anatomy speaks about the structure of the tooth, its appearance and how it is classified. If the subject concerns the structure and categories of the tooth, then this is described as macroscopic anatomy. The portion above the enamel junction is called the realm and is the crown of the tooth. This crown appears clearly if an eruption occurs. There can be either one or many roots in a single tooth.

In light of present-day eating habits, most of us are prone to dental diseases [24]. This is to the result of poor healthcare, low standards of living, unhygienic conditions and practices, and the use of drugs, tobacco and alcohol. In most situations, however, these diseases could be prevented if regularly monitored. The monitoring of these diseases can also play a vital role in controlling diabetes and also cerebrovascular and cardiovascular diseases.

Deep learning is widely used for object classification mainly when large data sets are involved. Algorithms such as CNN and RNN are used to predict diseases related to tooth using various forms of images [5].

Tooth loss normally occurs for baby teeth; once they are replaced by human adult teeth the possibility of tooth loss due to some disease is much reduced. Sometimes this will occur as a result of diseases such as dental avulsion, cavities, and gum disease. **Tooth plaque** is a sticky layer of bacterium that forms on teeth when we eat or drink. During this process, bacteria transforms the food into acids, which we term tooth plaque. The acids which are formed during this process destroy enamel.

Fluorosis is caused when teeth are overexposed to Halide (Halogen ion) [6]. This condition normally takes place during the initial eight years of childhood. It's a serious condition which affects the teeth. The normal result is that the teeth will lose color as a result. **Periodontal Disease** is to the result of poor oral hygiene. This leads to infection in the gums and affects the bone of the jaw. This can be stopped. In serious cases, it leads to tooth loss. This, in turn, can lead to respiratory issues [23]. Professionals may even perform surgeries in the most serious cases. If not treated, dental diseases can lead to heart disease, stroke, diabetes, some cancers, Alzheimer's, osteoporosis and even some metastatic diseases.

"You Only Look Once" (YOLO) is used by [25] for object detection. Deep learning algorithms which use object detection are as follows. One is based on classification and the other on regression. The occurrence of objects show up more often when the region of interest is identified which is proven in the classification algorithm [27]. When CNN is applied on the region of interest, the chances of object occurrence are high [25].

In this the detector must be run each time in the region of interest and this slows down the computation. This is a major drawback of this algorithm. Examples of such algorithms are R-CNN. The regression-based algorithms are faster when compared to this. In this algorithm, there is no choice of attention-grabbing ROI among the images [27]. Here, it considers the entire image and identifies the bounding boxes, thereby predicting the classes to which they belong. YOLO is one such regression-based algorithm. The YOLO detector is extraordinarily quick, hence its use in many applications, such as self-driving cars and others where period object detection is needed [25]. This chapter applies YOLO to identify the various tooth-related disorders. It is structured as follows. Section 14.2 presents the literature review of our work. Section 14.3 explains the methodology used in predicting the various tooth diseases using YOLO. In Section 14.4, the results are discussed. Finally, in Section 14.5 we give our conclusion and outline future work.

14.2 LITERATURE SURVEY

A literature survey or review is the section which shows the varied [1] analysis and research made within the field of our interest. In [22] author classified dental diseases using Convolutional Neural Networks. Diseases such as periapical infection, periodontitis and cavity were considered [21]. In order to obtain improved accuracy, transfer learning with a VGG16 pre-trained model is used. Training and testing was carried ot on 251 RVG x-ray images. A general accuracy of 88.46% was achieved. The limitations of this work are that: For training purposes, the system requires large labelled data sets; only when transfer learning was applied was an increase in accuracy identified; only x-ray images were used for classification of diseases; the length of the prediction time was not addressed; finally, the system seems to be more appropriate for use in a laboratory setting rather than being a suitable product for ordinary people [2].

The classification of skin was done using traditional methods wherein features were extracted manually. Later, in [18] the authors have applied CNN to extract features. The advantage of deep nets is that they need lesser pre-processing when compared with traditional methods [7]. Traditional methods have used various clustering techniques and Support Vector Machine for feature extraction. By contrast, better architectures such as deep nets have the ability to analyse images better.

Research has shown that only a comparatively small proportion of those people with tooth problems such as decay and fluorosis and so on visit a dentist. This is because of the lack of awareness among the population [31]. Hence awareness programs regarding oral health have to be conducted at regular intervals for the early diagnosis of diseases and the prevention of dental problems. Deep Convolution Neural Networks (CNN) were applied to medical problems with an input data size of over 10,000 images, which require [11] higher cognitive process. They were able to achieve an efficiency of 84%. Exploiting network cascades is a general task in the multitasking environment. The authors have shown that [4] these networks are fast and accurate during segmentation. Another framework was introduced for detecting objects deep [32] in the network. This framework is completely different from bounding box or sliding windows. Here, the corner point and the middle point are linked to the detection of objects. Hence, scale or ratio variation will not have any effect on this system.

IoT could be used for faster and cheaper [14] innovations for customers in healthcare. Diagnosis and location of intervertebral discs was done using filets and stacked autoencoders [16]. The data set for validation contained 102 images. Features were extracted using unsupervised learning techniques and the results, when compared, were reasonably good. Studies have shown that people with diabetes [19] and those with high levels of glycaemia are at a higher risk of being diagnosed with periodontitis. Other diseases, like cardiovascular and macrovascular complications, could also arise in patients with periodontitis. The Retinex algorithm was used [20] to scan the image in order to study the serology of an infection caused by pathogens. Various regression techniques were used in [29] to extract features to diagnose AD/MCI through neuroimaging. Dimensionality was a major problem in this approach. To overcome this issue, regression methods were used for feature selection, thereby reducing the dimensionality.

A survey was conducted by researchers in [30] where they have listed the various opportunities and research issues present in healthcare. A framework was used in [33] to demonstrate the optimization of general purpose algorithms. The authors have compared the various NFL theorems, to show how they underperform when the data set is small. AIDS patients generally have a low count of CD4 lymphocyte. Such patients are more prone to IRIS if they have had [10] an infection and have undergone treatment for the same. Such patients can be given treatment before HAART is initiated.

14.3 SUGGESTED DESIGN

The objective of this work is to train the images using YOLO3, to identify the tooth and the type of image (Figure 14.1) and then, later, to classify it into a proper class. The end product will be an app stored on the smartphone or tablet. YOLO is designed in such a way that at a time it will try to focus on each and every object which is available inside the images. In the faster R-CNN [10] family we try to send a feature which we will extract from the CNN and we will append the feature map with our proposed region.

Suppose there is an input image (Figure 14.4). We will try to send the input image with multiple objects in a pre-trained network (CNN). After

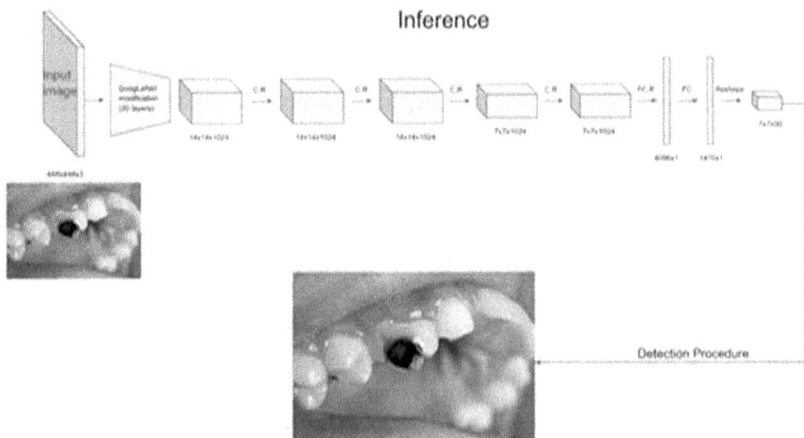

Figure 14.1 Tooth detection procedure.

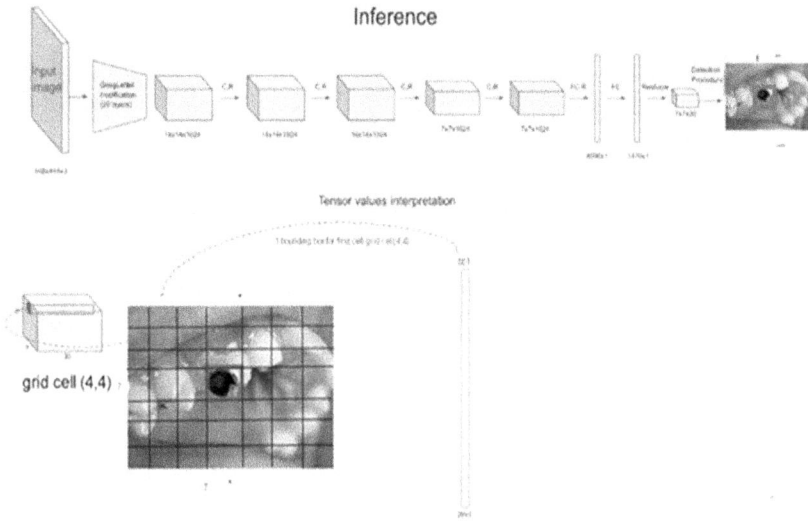

Figure 14.2 YOLO grid system.

sending the data it will try to give you different feature maps from different layers and send them into forward connection. Based on the previous feature map, we extract different objects in an input image.

The entire image will be split into grids (SxS) (Figure 14.2) where every box will have its own centre point (centroid). With respect to every box we are going to have two bounding boxes, which can have both vertical and horizontal coordinates (X,Y).

With respect to the image in Figure 14.2, we have a single class where every bounding box will have a width (W), a height (H) and confidence (C).

The model [34] then calculates the class score of class in each grid, leading to the output of each grid being 5 × 1; which means 5 is the number of classes and 1 is the grid. For each bounding box in each grid the process is repeated, leading this way class confidence of all 98 (7(X-axis) × 7(X-axis) × 2(1 + 1 1 on the X-axis and the Y-axis) grids in the image is generated.

Whenever the algorithm performs the first label of operation then it will be able to get the bounding boxes in the first layer. We then have to set the threshold for bounding boxes for the sum of the class as in Figure 14.3. If the result is less than the set threshold, then it chooses the next layer, and all the scores will be arranged in descending order [the Higher the confidence-First it comes [9]. YOLO uses NMS(Non maximal suppression algorithm). This eliminates all the unnecessary bounding boxes and keeps the boxes which are required and then it sets scores to zero for the redundant box.

In every bounding box we have 1-Class decay. The algorithm tries to extract confidence of decay of each and every bounding box (Figure 14.6) and it then takes the confidence score from each and every bounding box1

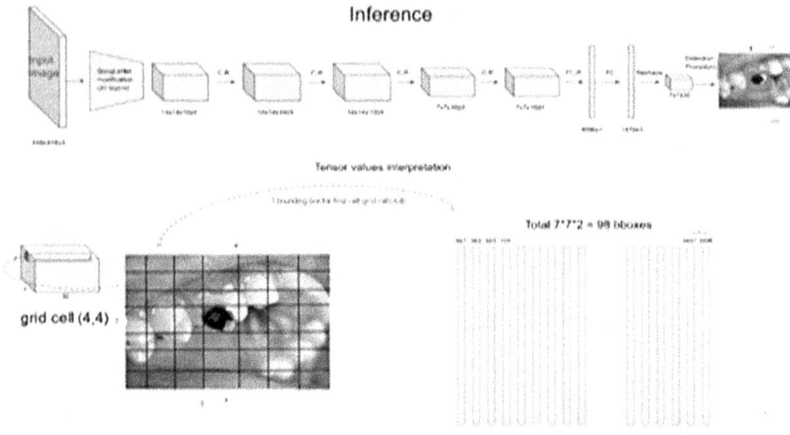

Figure 14.3 YOLO's bounding box plotting.

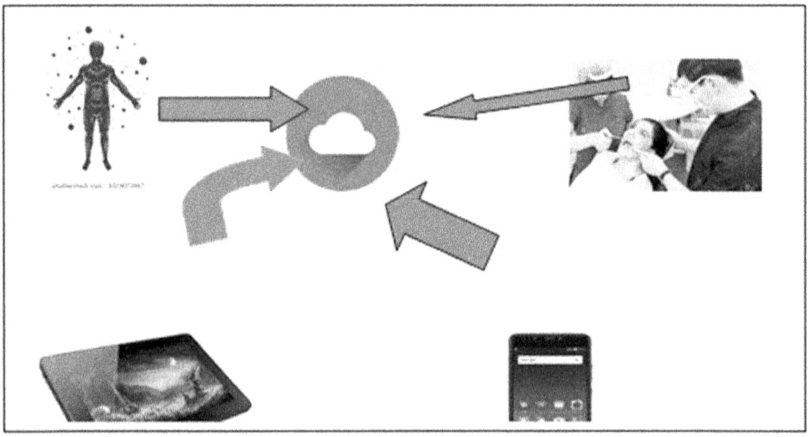

Figure 14.4 IoT Web cloud.

class i.e decay and arranges it in descending order. After arranging the boxes, it compares the top 2 scores as in Figure 14.5.

The comparison can be made by the equation given below

$$\text{IOU}(\text{box1})(\text{box2}) > 0.5 \text{ then set 0 score to } (\text{box2})$$

After evaluating all the bounding box scores, the value which is most accurate will be mapped and the output will be predicted. The classes included for training are fluorosis, decay, oral thrush, chipped tooth and fillings. For training of the model on Dental images [21], a pre-trained YOLO model,

Class Decay score for bounding box

Class Decay

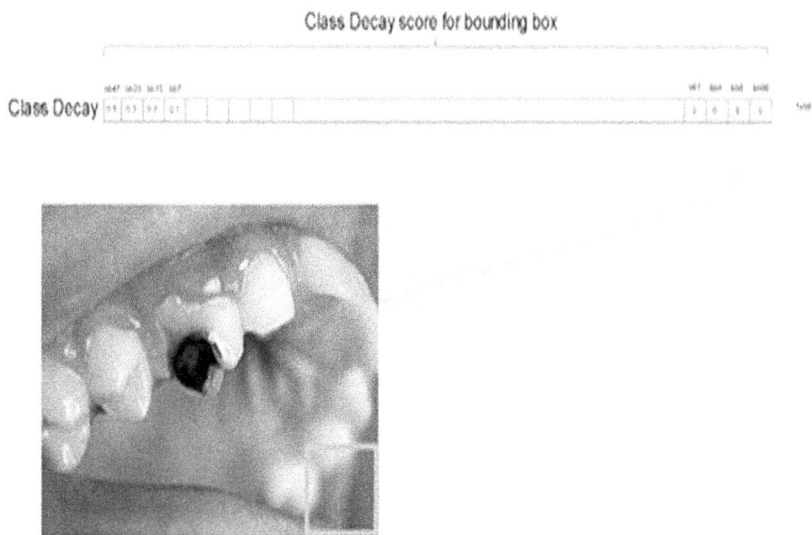

Figure 14.5 Comparison of scores.

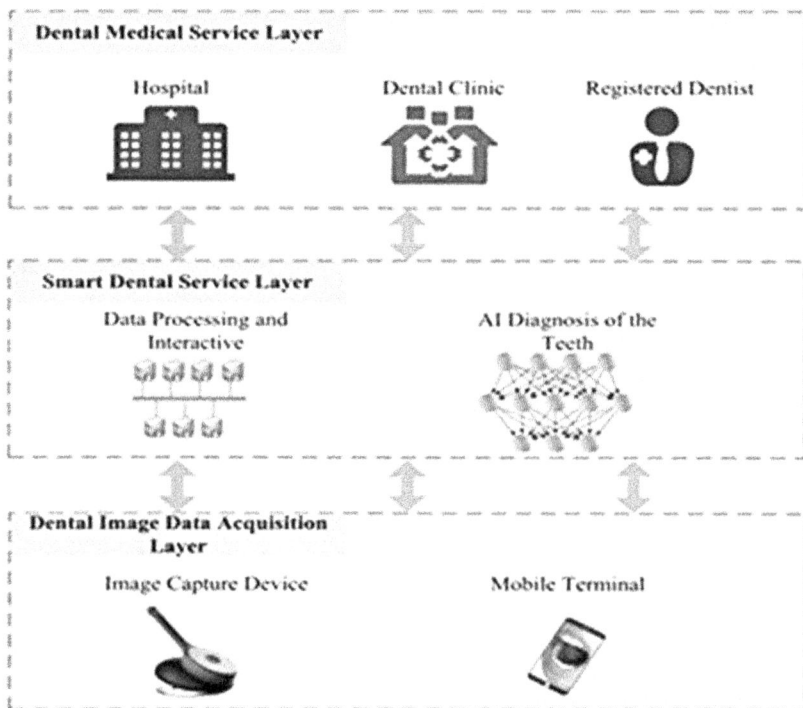

Figure 14.6 System network architecture for smart dental health SaaS system.

trained on the COCA and VOC data set, is used to improve the accuracy in achieving multi-object detection.

Optimizer Initialization: For the case study, an Adam Optimizer is being used. Adam is efficient when working with large problems which have many features. The combination of 'gradient descent with momentum' algorithm and the 'RMSP' algorithm results in this optimizer [1].

Warm-up steps are performed as it helps the network to slowly adapt to the data. However, theoretically, the main reason for warm-up steps is to allow adaptive optimizers (e.g. Adam, RMSProp, ...) to compute correct statistics of the gradients.

Three layers encompass the network architecture of the Smart Dental Health SaaS system shown in Figure 14.7

1. Input Image Layer
2. Intelligent Service Layer
3. Service Layer.

The outermost layer is the service layer, which coordinates with the registered dentists, healthcare centers, health providers and product suppliers. These centers provide attractive offers and attract customers through their infrastructure and technology.

This architecture helps in building positive loops with patients. This can be accomplished by managing the patients' data, recognize groups of patients who have shown considerable improvement in health conditions and informing them of the same, thereby creating a positive environment.

The core layer is called Smart Dental Service Layer, where the analysis results of dental symptoms are obtained by data processing [2, 28].

The middleware acts as an intermediate layer for the API. The API interacts with the middleware. All user request and logic support are provided by the cloud APIs.

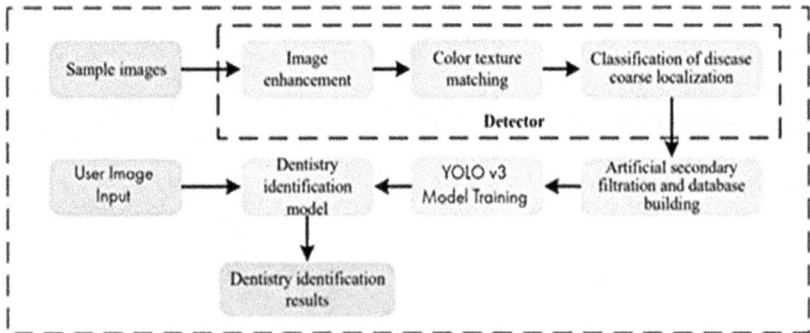

Figure 14.7 Dental image system logical data flow diagram.

The images are captured using a high-end camera or mobile. These devices do the local computing and display results to the end user through 3G, 4G or Wi-Fi networks. The service layer assists users in checking the health condition of their tooth. The data captured by these devices is fed to the service layer. YOLO algorithm is then applied on these images to identify the area of infection, if any. If the service layer confirms any defect in the tooth, then medical advice may have to be taken. The dental images are stored using Mongo database. The flow of services is shown in Figure 14.10. The application has three layers: the input layer, the service layer and the client layer.

HTTP service is accomplished using allocation and scheduling resources through load-balancing techniques. It also helps in implementing the connections between the algorithms and the cloud. The logical flow diagram is shown in Figure 14.7.

Steps in the flow diagram are:

1. Sample images are acquired and fed into the system.
2. The sample images are enhanced for accurate and common ground predictions.
3. The colour texture is matched on to a single scale.
4. Classification of disease and coarse localization is done by using MS VoTT software, where disease labeling and ROI (Region of Interest) are determined.
5. Further preparation like train and test split of images and model weights are added and segregated for model training.
6. The data is fed into YOLO and trained on training images [8].
7. The trained model is then loaded on to the application software for consumption
8. The user then uploads the picture to the trained and validated model.
9. Predictions are made and the image is sent back to the user

14.3.1 YOLO object detection interpretation

Unlike the RCNN series, YOLO (Figure 14.12) classifies all problems related to detection as a regression problem. It uses box regression and network for classification [3, 12]. The specific method is: divide the image into N * N grids, where the function of the grid is to predict the positions (x, y, w, h) of Yb boxes, confidence, and class probability. The output dimension is N * N * (Y * 5 + Ca), where Ca indicates the number of categories. The responsibility of each grid is to predict a set of class probabilities. The number of grids will not have any impact on the class probabilities.

1. Prediction of the Bounding box
   YOLO foretells that anchor boxes are used by bounding boxes (Figure 14.12). This network figures 4 coordinates for each bounding box: ax,

ay, aw, and ah. (tx, ty) represents [28] the coordinates of the upper-left corner of the current grid offset from the upper-left corner of the image and it is the distance, with pw, ph representing the width and height of the prior box (prior). During training, a sum of squared error loss is used. True value by said gradient is the true value minus the predicted value. YOLO uses logistic regression to predict a score for each bounding box. The algorithm only matches an optimal prior box for each truth value.

2. Multi-label prediction
Each bounding box may contain multiple types of objects, that is, multi-label prediction. So logistic (sigmoid) is used instead of Soft-Max, because SoftMax indicates that each box has only one type of object, and sigmoid can predict multi-label classification. In fact, a [35] sigmoid classifier is applied to the results of each classification calculation in logits to determine whether the sample belongs to a certain category. During training, a binary cross-entropy loss function is used for class prediction [13].

3. Combine the characteristics of different convolution layers to extract more fine-grained information and make multi-scale predictions
YOLO uses three different scales to predict the boxes. The network uses a similar feature pyramid concept to extract features from different scales. Several new convolution layers will be added to the original basic feature extractor. Finally, a 3-dimensional tensor will be used to represent the bounding box, objectness, and class predictions. The next step is to predict 3 boxes, so the tensor is N * N * [3* (4 + 1 + 80)]. This represents 80 class predictions,1 objectness prediction and 4 bounding boxes. Then we retrieve a feature set from the previous network and merge it with the up sampled features. The objective is to get an approximation of what would have been obtained if it was done at a higher rate. The next step is to append some additional convolutional layers so that an alike tensor could be produced by processing this feature map. Repeat this process to predict the size of the final box. In the past, YOLO had difficulty in predicting small objects, but now it has improved significantly through multi-scale prediction.

4. Network structure (DarkNet53 = Darknet19 + ResNet)
Combine the residual thought to extract deeper semantic information. Continuous 3 × 3 and 1 × 1 convolutional layers are still used. Prediction will be performed on three different scales by up sampling. For example, the 8 * 8 feature map up sampling and the 16 * 16 feature maps are added and calculated again, so that smaller objects can be predicted. A convolutional layer with a step size of 2 is used instead of the pooling layer because the pooling layer will lose information.

5. Predict more targets
YOLO still uses k-means clustering to determine template boxes and predict 9 bounding boxes for each grid, which can improve recall.

6. Loss function

Following the reference from [24], performance can be improved if we add pre-trained convolutional networks. Input resolution is increased to 448*448 from 224*224. This is done because visual images which are fed should be fine-grained so that object detection becomes accurate. The network consists of four convolution layers, two of which are fully connected with weights initialized at random. The next step is to compress the values in between 0 and 1. This can be achieved by normalizing the width and height of the bounding box image. Next parameterize the bounding box x and y coordinates to be the offsets of a particular grid cell location so they are also bounded between 0 and 1. Leaky rectified linear activation function is applied on the intermediate layers and a linear activation function is used for the final layer.

$$\phi(x) = \begin{cases} x, \text{if } x > 0 \\ 0.1x, \text{otherwise} \end{cases}$$

The next step is to get an optimum model by finding out the sum of squared error Figure 14.8. This results in zero confidence score and high loss [15]. In order to reduce the loss, we use two parameters, λ_{coord} and λ_{noobj}. These parameters are used to determine the error based on the deviations. If the

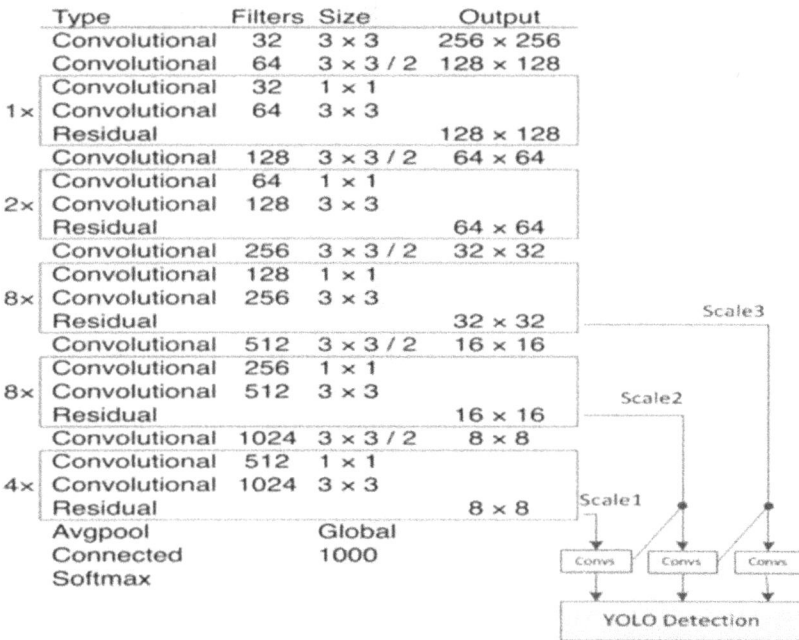

Figure 14.8 YOLO detection using convolution.

deviations are large, they cause greater effect in small boxes rather than the large ones. To overcome this, the square root of the bounding boxes height and width are considered instead of using them directly.

$$\gamma_{cod} \sum_{x=0}^{n^2} \sum_{y=0}^{M} 1_{ij}^{obj} \left[\left(a - a_{\hat{i}} \right)^2 + \left(b_i - b_{\hat{i}}^{\wedge} \right)^2 \right]$$

$$+ \gamma_{coord} \sum_{x=0}^{n^2} \sum_{y=0}^{M} 1_{ij}^{obj} \left[\left(\sqrt{c_i} - \sqrt{c_{\hat{i}}^{\wedge}} \right)^2 + \left(\sqrt{h_i} - \sqrt{h_{\hat{i}}^{\wedge}} \right)^2 \right]$$

$$+ \sum_{x=0}^{n^2} \sum_{y=0}^{M} 1_{ij}^{obj} \left[\left(C_i - C_{\hat{i}}^{\wedge} \right)^2 \right] + \lambda_{noobj} \sum_{i=0}^{s^2} \sum_{j=0}^{B} 1_{ij}^{obj} \left[\left(C_i - C_{\hat{i}}^{\wedge} \right)^2 \right]$$

$$+ \sum_{i=0}^{s^2} 1_{ij}^{obj} \sum_{c \in \text{Classes}} \left(p_i(c) - p_{\hat{i}}(c) \right)^2$$

During the training phase, for each grid cell, the algorithm predicts a number of bounding boxes. Each object is overlooked by one predictor during the training phase. This predictor later predicts the objects based on IOU. The one which has the highest IOU w.r.t ground truth will be the prediction. With this the predictors get specialized. The overall recall starts improving as predictions are able to predict the aspect ratios, sizes or various classes of objects better. The assumptions made during training are as follows. The object which is in cell i is denoted as 1_{ij}^{obj}. It also says that the prediction is due to the jth bounding box present in cell i. Another fact observed here is that the penalization of classification error happens by the log function only if the grid cell contains that particular object. In the case that the predictor has the greatest IOU of any grid cell, then the bounding box coordinate gets penalized.

14.4 OVERALL NETWORK STRUCTURE

The basic network model is Google Net, but instead of using the inception module, we have used 1 * 1 and 3 * 3 convolutional layers alternately. Convolutional layer extracts features from the fully connected layer, a total of 24 convolutional layers, 2 fully connected layers.

The network consists of 24-Convolutional + 2 fully connected layers [35].

Sub-network: This is a pre-trained classification network, the input image size is 224 * 224. It contains first 20 convolutional layers + 1 global average pooling + 1 fully connected. Sub-network: The second one is a target detection network, the input image size is 448 * 448. Loss function (square sum loss function) including 4 parts: box center position x, y loss + box width and height w, h loss + confidence loss + classification loss.

YOLO advantages

- High speed. Seen as a regression problem, no complicated pipeline is needed. Have a global understanding of the image.
- Other than RCNN, the features of the whole image is considered to predict the bounding box, instead of RCNN, only the features of the candidate bounding box can be used to predict the box.

The number of candidate boxes is much smaller, only 7 * 7 * 2 = 98. The RCNN has 2000 selective searches, which are computationally intensive. Representation of Bounding boxes is given in Figure 14.10.

In our model (in Figure 14.9), YOLO has used dimension clusters as anchor boxes to predict bounding boxes. Let us assume that the cell is at an offset from the top left corner of the image by a dimension (c_x, c_y) and the width and height of the bounding box is p_w, p_h.

$$b_x = \sigma t_x + c_x \tag{14.1}$$

$$b_y = \sigma t_y + c_{xy} \tag{14.2}$$

$$b_w = p_w e^{t_w.} \tag{14.3}$$

$$b_h = p_h e^{t_h} \tag{14.4}$$

The performance parameters used to compute loss during training is the sum of the squared error. The gradient is the error term computed as the difference between ground truth value i.e. t^* and prediction which is t^* − t*. If the equations above are inverted, it displays the ground truth value. To predict the objectness score for each bounding box, YOLO uses logistic regression. The score should have a value of 1 if the bounding box

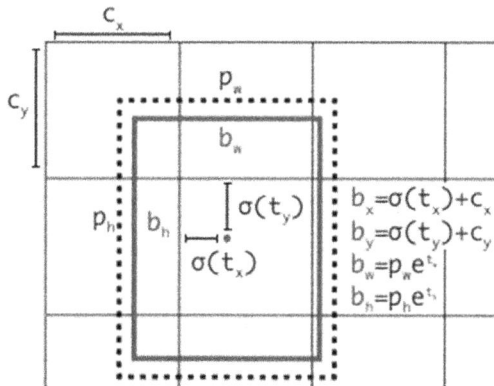

Figure 14.9 Bounding boxes.

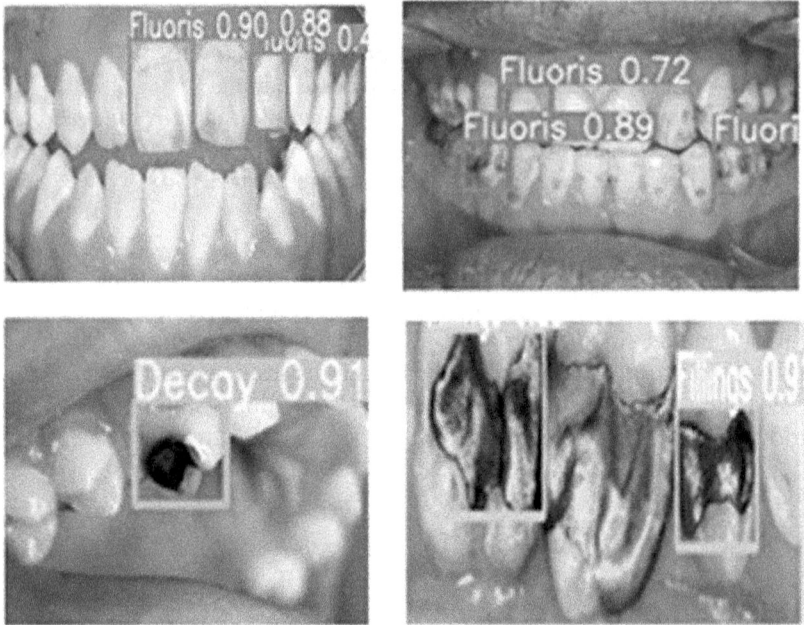

Figure 14.10 Bounding box illustration.

prior overlaps a ground truth object by more than any other bounding box prior. The cluster centroids can be used to estimate the width and height of the box.

The sigmoidal function is used to predict the centre coordinates of the box relative to the location of the filter. The value of threshold is 0.5. Each ground truth object is assigned one bounding box prior. Sometimes there will be no loss incurred if the assignment between bounding box prior and ground truth doesn't happen.

Bounding boxes (Figure 14.10) are boxes that are imaginary boxes. They are around the objects checked for collision. There are two types of coordinate systems, a 2D coordinate system and a 3D coordinate system, that are used to detect objects. In Figure 14.10, the square boxes represent the bounding boxes which, in turn, show the affected disease part of the tooth. The accuracy of the prediction is also shown along with the bounding boxes. In Figure 14.13, the digits along with the bounding boxes 0.90, 0.88, 0.72, 0.89 etc. represents accuracy of 90%, 88%, 72% and 89%

14.5 RESULTS AND DISCUSSION

Figures 14.11–14.15 show how the trained model is predicting dental disease.

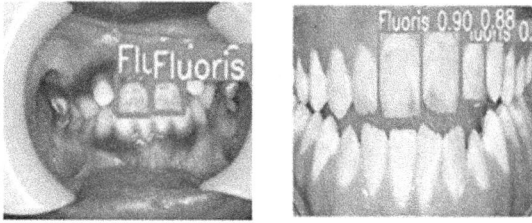

Figure 14.11 Predicted results of dental disease fluorosis.

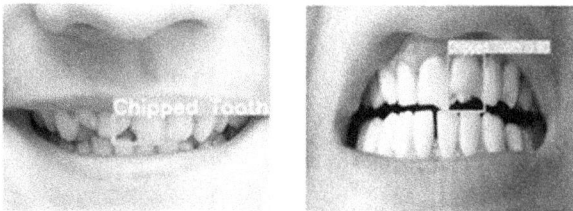

Figure 14.12 Predicted results of dental disease chipped tooth.

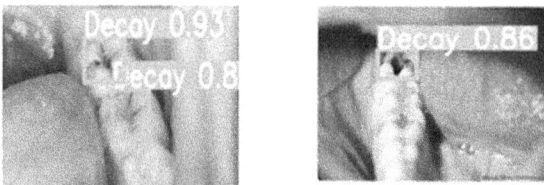

Figure 14.13 Predicted results of dental decay.

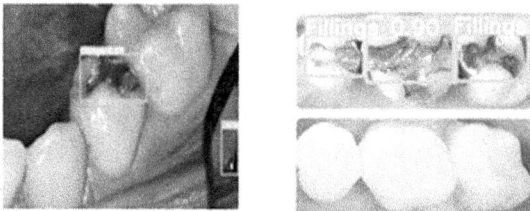

Figure 14.14 Predicted results of dental disease fillings.

The predicted model gives the results with predicting bounding boxes with accuracy mentioned over it. Diseases like fluorosis, chipped tooth, tooth decay, oral thrush and fillings are predicted by the proposed system. If the value of the bounding box is 0.9 then this says that it is 90% accurate.

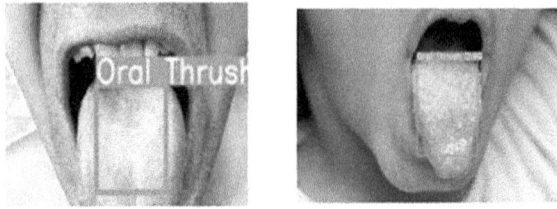

Figure 14.15 Predicted results of dental disease oral thrush.

14.5.1 Performance measure

The metrics used for calculation of performance measures are IoU and Precision and Recall. The same is illustrated in Figure 14.17.

14.5.2 IoU (Intersection Over Union)

- In object detection models, IoU is used to evaluate the performance of the model.
- IoU (Figure 14.16) is the division of the union of two bounding boxes for the ground truth to the predicted bounding box.

14.5.3 Precision and recall curve

Precision: It shows how many times the model is correct in positive prediction. It gives an idea how much we can trust our algorithm when it predicts "true". The proportion of predictions guessed properly as "true" versus all the positive predictions.

Recall: Recall tells us the percentage of all the correct output classified by the model.

If we want to know the best threshold value of our proposed system then we have to plot a Precision Recall curve (PR curve). Therefore, we have to

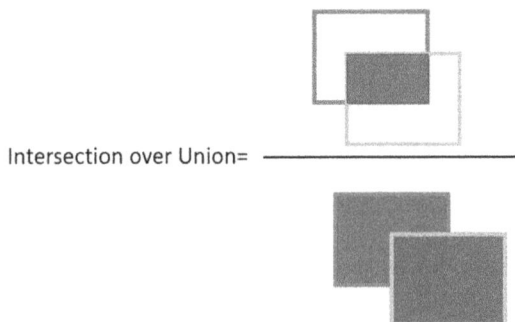

Intersection over Union=

Figure 14.16 Representation of intersection over union.

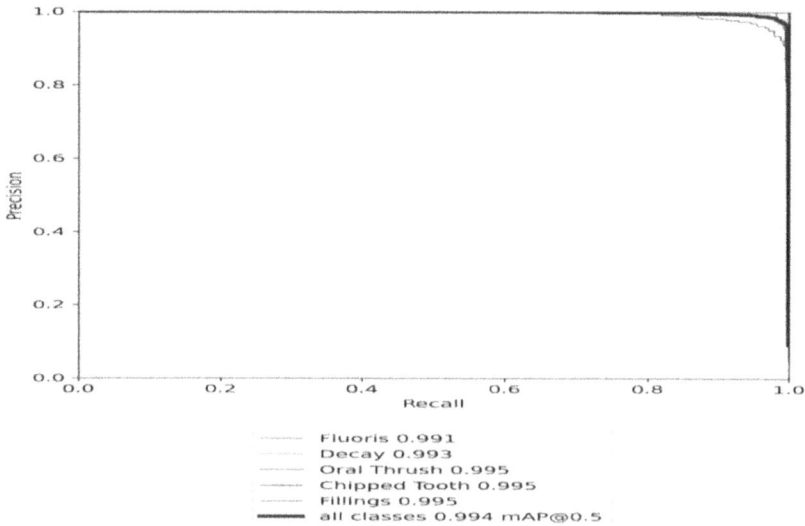

Figure 14.17 Precision and recall curve.

calculate the precision and recall values of the algorithm for different values of threshold. We then use those values to plot a graph. In Figure 14.17, the blue curve is the best for our proposed system.

In Figures 14.17–14.18, the curve of Precision and Recall is drawn. We have obtained mAP (mean Average Precision) of the respective diseases. It is evident that the mAP value is maximum for three diseases i.e. oral thrush, chipped tooth, fillings is about 99.5% and fluorosis is of 99.1% and tooth decay is of 99.3% respectively.

The higher the mAP value the greater the accuracy in respective disease prediction. We can conclude that the average mAP value of all diseases is about 99.4%.

An 80% confidence interval means that we are confident at 80% that the real value is in the specified interval. Large samples output mean with more precision when compared with smaller data. As a result, when a large sample is taken, the confidence could be smaller. As we increase the threshold of confidence the precision will go up (Figure 14.19). The average confidence of all the classes in the above graph is 86% at 1.0.

Using the F1 score curve shown in Figure 14.20, the balance between precision and recall can be visualized and a design point can be determined using a curve graph. The F1 curve shows that optimal value for precision and recall is obtained when confidence is 0.506. We would want our model to have cases which have higher confidence value. The maximum value of F1 occurs at 0.81. Hence it is ideal to select confidence of 0.79 as it results in F1 value of 0.8 which is pretty close to the maximum F1 value.

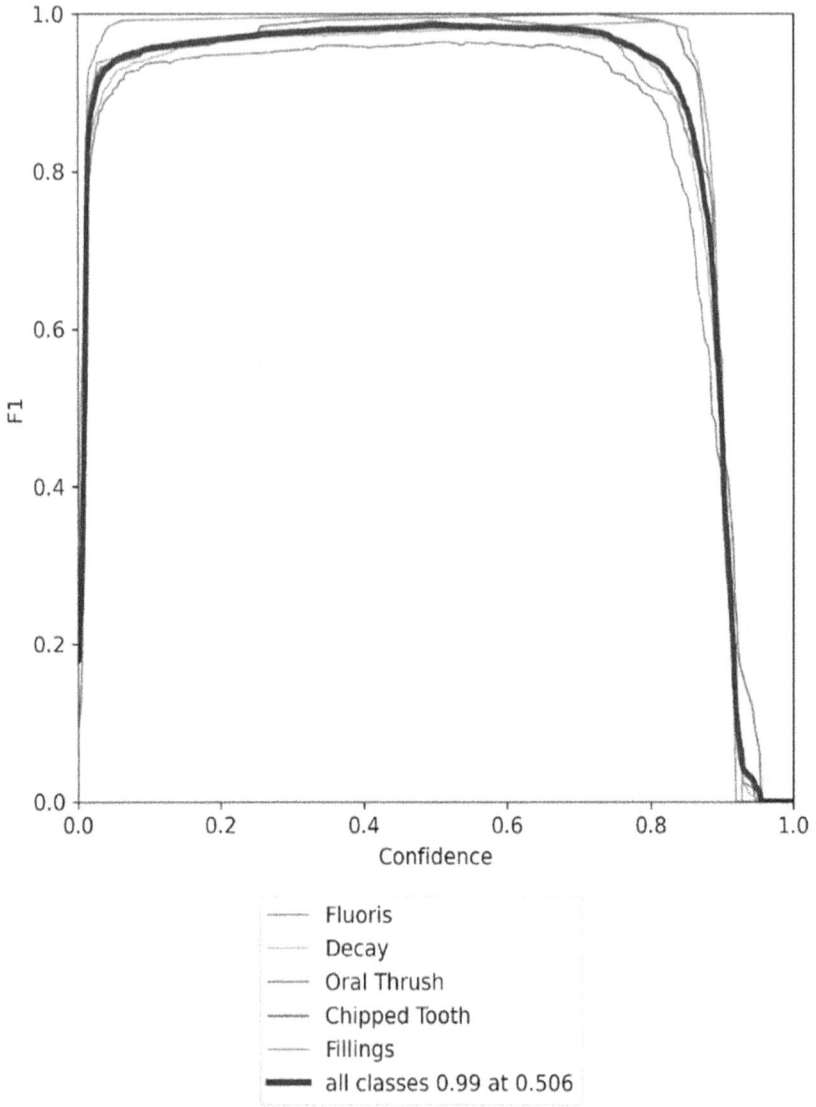

Figure 14.18 F1 and confidence curve.

14.5.3.1 Analysis of bounding boxes using histogram

To describe the spatial location of an object we use a bounding box in object detection.

The bounding box is usually rectangular in shape where it has x- and y-coordinates. Another commonly used bounding box representation is the

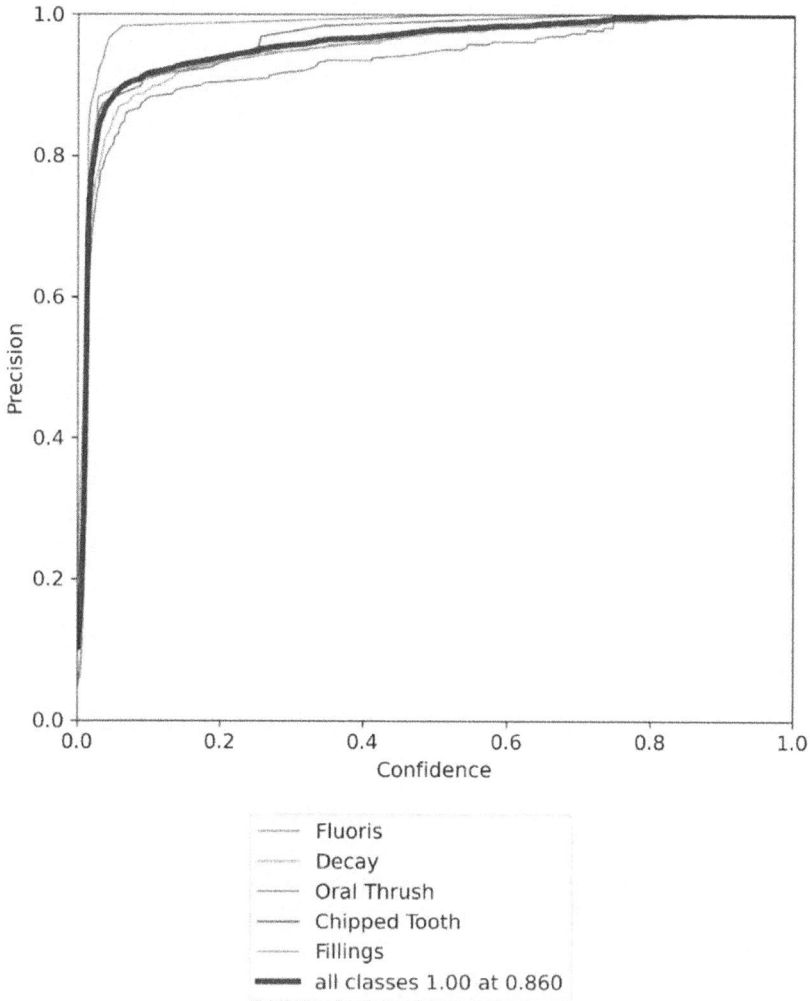

Figure 14.19 Precision and confidence curve.

(x, y) (x, y)-axis coordinates of the bounding box center, and the width and height of the box, as shown in the histogram in Figure 14.21.

14.5.3.2 Analysis of bounding boxes using different type pf graphs

A bounding box has four values to represent it, i.e. x_center, y_center, width, height. The center of the bounding box is x_center and y_center is the normalized coordinates.

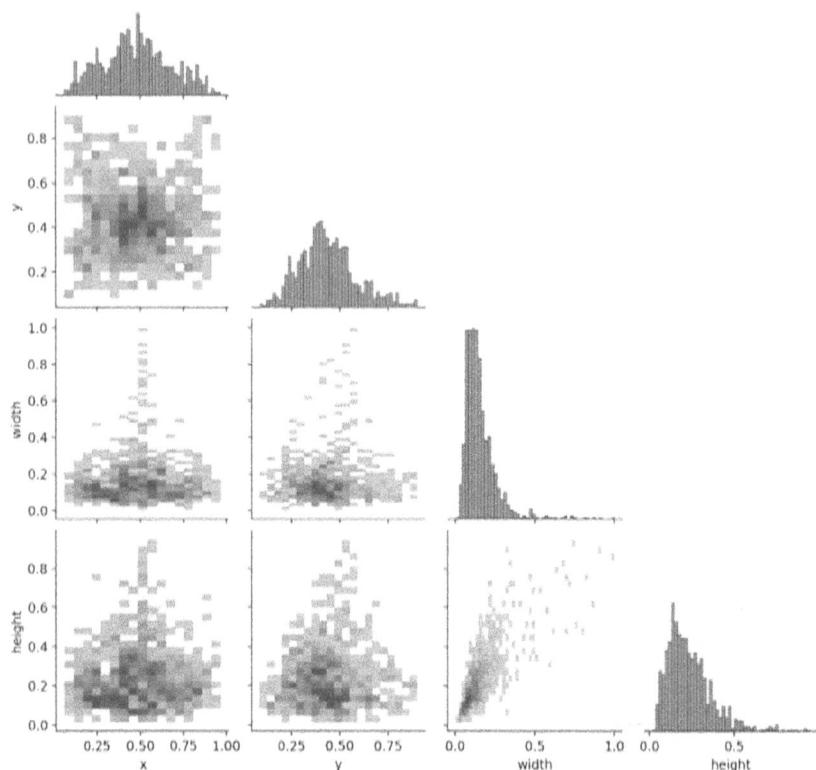

Figure 14.20 Histogram plotting for bounding boxes.

We consider pixel values of x and y to normalize the coordinates. This becomes the center of the bounding box on the x- and y-axes. Next, the value of x is divided by the width of the image and value of y is divided by the height of the image. They are also normalized.

In Figure 14.24, we can see the different types of representation of the bounding boxes. Each graph takes two parameters (either x and y) or height and width to represent the bounding boxes of the dental data. We can also see the range of instances of each disease by representing it in the bar graph. The range covered by all the bounding boxes in the data set is also shown as a summary.

14.5.3.3 Metrics analysis

The graphs in Figures 14.22–14.24 show the performance of each metric of the YOLO algorithm.

Recall: This parameter indicates the number of correct instances that were [26] obtained.

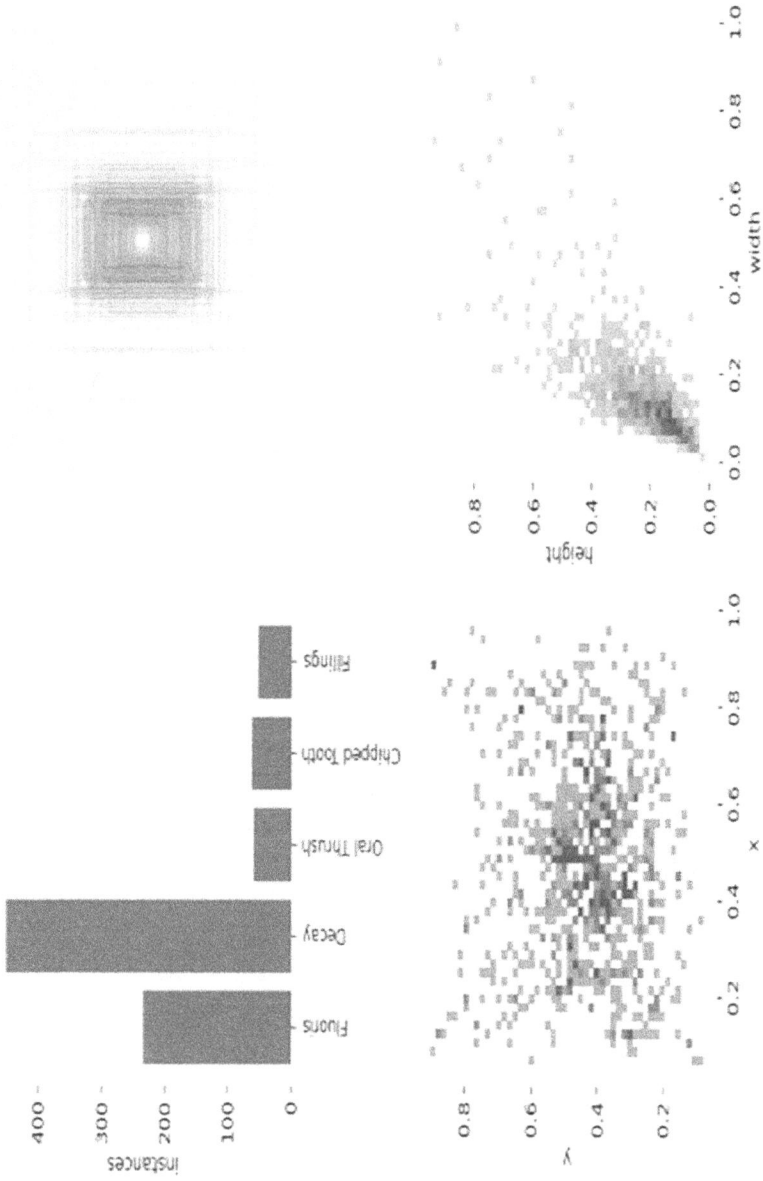

Figure 14.21 Different graph plotting for bounding boxes.

metrics/recall

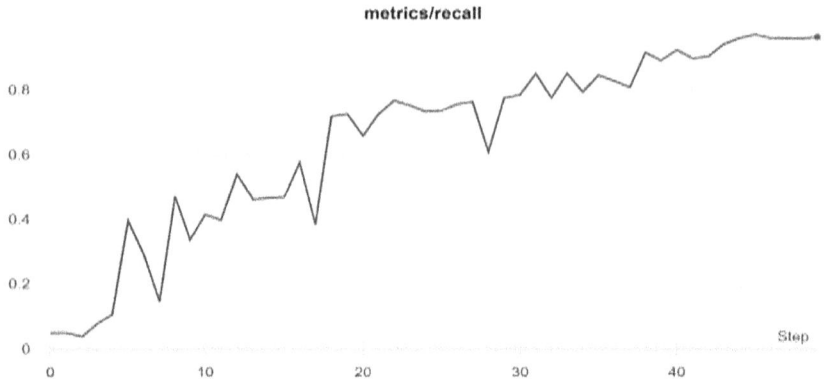

Figure 14.22 Performance of recall.

metrics/mAP_0.5

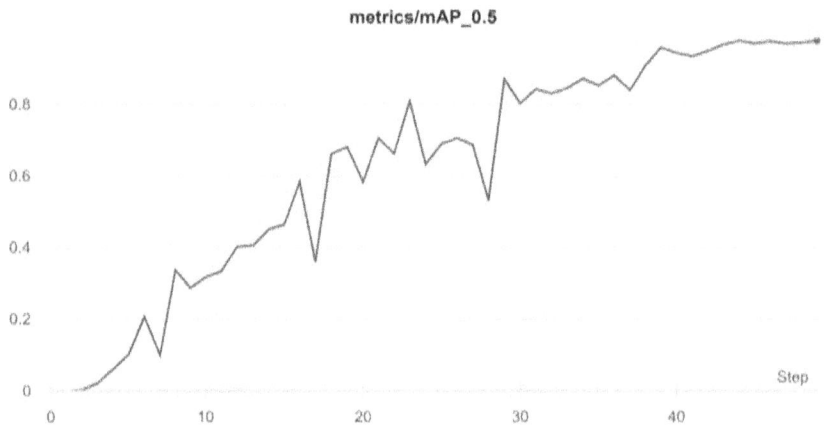

Figure 14.23 Performance of metric mAP.

mAP: To measure the accuracy of object detectors, the metric used is Average [17] Precision.

In this we try to find the area covered by the precision recall curve. Since the values of recall and precision lie between 0 and 1, the value of mAP will also lie between 0 and 1.

$$AP = \int_0^1 pr(re)dr \qquad (14.5)$$

The x-axis represents the epochs count and the y-axis represents the respective metric values evaluated during the training of the data set.

metrics/precision

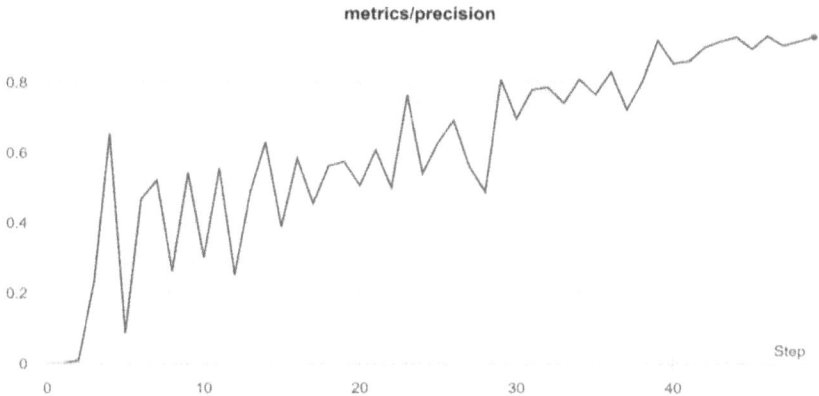

Figure 14.24 Performance of metric precision.

The recall, precision and mAP gradually increase with the increase in epochs count. It further leads to an increase in the accuracy of the overall model.

14.6 CONCLUSION

The present results demonstrates the prediction of tooth diseases such as tooth decay, tooth filling, fluorosis, chipped tooth and oral thrush through an app. We have seen that YOLO is very good at object detection and produces accurate results. Recall of 91.3% was obtained, as shown in the graph. Precision of around 89% was obtained with 45 step counts. We have observed that as the number of epochs was increased, the accuracy used to increase. This study has its own limitations, one of them being that a high-speed network is necessary for the app to work well. At times, dentists may have to perform some physical tests for the diagnosis of the disease which cannot be provided by our system. In addition to this, the app on the mobile can be used in the diagnosis and to evaluate the disease. Currently, many nations may not have such access to the technology which is required for our system to work. In future, we hope to overcome all these flaws and build a more robust system.

REFERENCES

1. Abar S, Abe T, Kinoshita T, "A next generation knowledge management system architecture," In *18th international conference on advanced information networking and applications, AINA 2004*, (vol. 2, pp. 191–195). IEEE, 2004.
2. Birogul S, Temür G, Kose U, "YOLO object recognition algorithm and buy-sell decision model over 2D candlestick charts," *IEEE Access*, vol. 8, pp. 91894–91915, 2020.

3. Chen, CH, *Handbook of pattern recognition and computer vision*, World Scientific, 2015.
4. Dai J, He K, Sun J, "Instance-aware semantic segmentation via multi-task network cascades," In *Proceedings of the IEEE conference on computer vision and pattern recognition*. pp. 3150–3158, 2018. doi: 10.1109/CVPR.2016.343
5. Esteva A, Kuprel B, Novoa RA, Ko J, Swetter SM, Blau HM, Thrun S, "Dermatologist-level classification of skin cancer with deep neural networks," *Nature*, vol. 542, no. 7639, pp. 115–118, 2017.
6. He K, Zhang X, Ren S, Sun J, "Deep residual learning for image recognition," In *Proceedings of the IEEE conference on computer vision and pattern recognition*. pp. 770–778, 2016. doi: 10.1109/CVPR.2016.90
7. Henriques J, Neves N, de Carvalho P, "XV Mediterranean Conference on Medical and Biological Engineering and Computing," In *MEDICON Proceedings of MEDICON 2019*, Coimbra, Portugal, vol. 76, Springer Nature, September 26–28, 2019.
8. Hua, G., Jégou H, "Computer Vision–ECCV", 2016 Workshops: Amsterdam, The Netherlands, 2016, Proceedings, Part II (Vol. 9914). Springer, 2016.
9. Ibadov, SR, Kalmykov, BY, Ibadov RR, Sizyakin, RA, " Method of automated detection of traffic violation with a convolutional neural network," In *EPJ Web of Conferences*, vol. 224, pp. 04004. EDP Sciences, 2019.
10. Khanagar SB, Alfouzan K, Awawdeh, M, Alkadi L, Albalawi F, Alfadley A, "Application and performance of artificial intelligence technology in detection, diagnosis and prediction of dental caries (DC)—A systematic review," *Diagnostics*, vol. 12, no. 5, p. 1083, 2022. doi: 10.3390/diagnostics12051083
11. Kwasigroch A, Mikołajczyk A, Grochowski M, "Deep convolutional neural networks as a decision support tool in medical problems–malignant melanoma case study," In *Polish Control Conference*, Springer, Cham, pp. 848–856, Jun 18, 2017.
12. Li, S, Gu, X, Xu, X, Xu, D, Zhang, T, Liu Z, Dong Q, "Detection of concealed cracks from ground penetrating radar images based on deep learning algorithm," *Construction and Building Materials*, vol. 273, pp. 121949, 2021.
13. Liu S, Li, X, Gao M, Cai Y, Nian R, Li P, Yan T, Lendasse A, "Embedded online fish detection and tracking system via YOLOv3 and parallel correlation filter," In *OCEANS 2018 MTS/IEEE Charleston*, pp. 1–6, IEEE, 2018, Oct.
14. Metcalf D, Milliard ST, Gomez M, Schwartz M, "Wearables and the internet of things for health: Wearable, interconnected devices promise more efficient and comprehensive health care," *IEEE Pulse*, vol. 7, no. 5, pp. 35–39, 2016. http://doi:10.1109/MPUL.2016.2592260.
15. Nie G, Liu Y and Wang Y, "DovCut: a draft based online video compositing system," In *Proceedings of the 15th ACM SIGGRAPH Conference on Virtual-Reality Continuum and Its Applications in Industry*, Volume 1, pp. 129–136, 2016 Dec.
16. Oktay AB, Akgul YS, "Diagnosis of degenerative intervertebral disc disease with deep networks and SVM," In *International Symposium on Computer and Information Sciences*, Springer, Cham. pp. 253–261, Oct 27, 2016.
17. pastel.archives-ouvertes.fr

18. Pang Z, Chen Q, Tian J, Zheng L, Dubrova E, "Ecosystem analysis in the design of open platform-based in-home healthcare terminals towards the internet-of-things. In *2013 15th international conference on advanced communications technology (ICACT)*, IEEE, pp. 529–534, Jan 27, 2013.

19. Preshaw PM, Alba AL, Herrera D, Jepsen S, Konstantinidis A, Makrilakis K, Taylor R, "Periodontitis and diabetes: a two-way relationship," *Diabetologia*, vol. 55, no.1, pp. 21–31, 2012. https://dx.doi.org/10.1007%2Fs00125-011-2342-y.

20. Provenzi E, De Carli L, Rizzi A, Marini D, "Mathematical definition and analysis of the Retinex algorithm," *JOSA A.*, vol. 22, no. 12, pp. 2613–2621, 2005.

21. Prajapati SA, Nagaraj R and Mitra S, "Classification of dental diseases using CNN and transfer learning," In *2017 5th International Symposium on Computational and Business Intelligence (ISCBI)*, IEEE, pp. 70–74, Aug, 2017.

22. Prajapati SA, Nagaraj R, Mitra S, "Classification of dental diseases using CNN and transfer learning," In *2017 5th International Symposium on Computational and Business Intelligence (ISCBI)*, IEEE, pp. 70–74, Aug 11, 2017.

23. Pussinen PJ, Jousilahti P, Alfthan G, Palosuo T, Asikainen S, Salomaa V, "Antibodies to periodontal pathogens are associated with coronary heart disease," *Arteriosclerosis, Thrombosis, and Vascular Biology*, vol. 23, no.7, pp. 1250–1254, 2003. http://dx.doi.org/10.1161/01.ATV.0000072969.71452.87.

24. Ren S, He K, Girshick R, Zhang X, Sun J, "Object detection networks on convolutional feature maps," *IEEE Transactions On Pattern Analysis and Machine Intelligence*, Aug 17, vol. 39, no. 7, pp. 1476–1481, 2016. http://dx.doi.org/10.1109/TPAMI.2016.2601099.

25. robocademy.com

26. Sathya, B. and Neelaveni, R., "Transfer learning based automatic human identification using dental traits-an aid to forensic odontology," *Journal of Forensic and Legal Medicine*, 76, pp. 102066, 2020. http://dx.doi.org/10.1016/j.jflm.2020.102066.

27. Smys, S., Tavares, J.M.R.S., Balas, V.E. and Iliyasu, A.M., "Computational Vision and Bio-Inspired Computing," 2019.

28. southern-sweepers.com.

29. Suk HI, Lee SW, Shen D, "Deep sparse multi-task learning for feature selection in Alzheimer's disease diagnosis," *Brain Structure and Function*, vol. 221, no. 5, pp. 2569–2587, 2016. http://doi:10.1007/s00429-015-1059-y.

30. Sundaravadivel P, Kougianos E, Mohanty SP, Ganapathiraju MK, "Everything you wanted to know about smart health care: Evaluating the different technologies and components of the internet of things for better health," *IEEE Consumer Electronics Magazine*, Dec 13, vol. 7, no. 1, pp. 18–28, 2017. https://doi.org/10.1109/MCE.2017.2755378.

31. Varghese CM, Jesija JS, Prasad JH, Pricilla RA, "Prevalence of oral diseases and risks to oral health in an urban community aged above 14 year," *Indian Journal of Dental Research*, 2019, vol. 30, no.6, pp.844, http://doi:10.4103/ijdr.IJDR_42_18.

32. Wang X, Chen K, Huang Z, Yao C, Liu W, "Point linking network for object detection," arXiv preprint arXiv:1706.03646. Jun 12, 2017.

33. Wolpert DH, Macready WG, "No free lunch theorems for optimization," *IEEE Transactions On Evolutionary Computation*, vol. 1, no. 1, pp. 67–82, 1997. http://doi.org.10.1109/4235.585893.

34. Almalki, YE, Din, AI, Ramzan, M, Irfan, M, Aamir, KM, Almalki, A, Alotaibi, S, Alaglan, G, Alshamrani, HA, Rahman, S, "Deep learning models for classification of dental diseases using orthopantomography X-ray OPG images", *Sensors*, vol. 22, no. 19, p. 7370, 2022.

35. Zhao, K. and Ren, X., "May. Small aircraft detection in remote sensing images based on YOLOv3", *IOP Conference Series: Materials Science and Engineering*, vol. 53, no. 1, pp. 012056, IOP Publishing, 2013.

Index

Pages in *italic* refer figures.

For Product Safety Concerns and Information please contact our EU
representative GPSR@taylorandfrancis.com
Taylor & Francis Verlag GmbH, Kaufingerstraße 24, 80331 München, Germany